ARCO

Everything You Need to Score High on

FIREFIGHTER

13TH EDITION

ARCO

Everything You Need to Score High on

FIREFIGHTER

13TH EDITION

EL DORADO COUNTY LIBRARY
345 FAIR LANE
PLACERVILLE, CA 95667

Fred M. Rafilson, Ph.D.
I/O Solutions, LLC
Illinois Fire and Police Recruitment Administration, Inc.

Robert Andriuolo
Former Deputy Chief, New York City Fire Department

ARCO
THOMSON LEARNING

Australia • Canada • Mexico • Singapore • Spain • United Kingdom • United States

An ARCO Book

ARCO is a registered trademark of Thomson Learning, Inc., and is used herein under license by Peterson's.

About Peterson's

Founded in 1966, Peterson's, a division of Thomson Learning, is the nation's largest and most respected provider of lifelong learning online resources, software, reference guides, and books. The Education Supersite^{SM} at petersons.com—the Web's most heavily traveled education resource—has searchable databases and interactive tools for contacting U.S.-accredited institutions and programs. CollegeQuest^{SM} (CollegeQuest.com) offers a complete solution for every step of the college decision-making process. GradAdvantage^{TM} (GradAdvantage.org), developed with Educational Testing Service, is the only electronic admissions service capable of sending official graduate test score reports with a candidate's online application. Peterson's serves more than 55 million education consumers annually.

Thomson Learning is among the world's leading providers of lifelong learning, serving the needs of individuals, learning institutions, and corporations with products and services for both traditional classrooms and for online learning. For more information about the products and services offered by Thomson Learning, please visit www.thomsonlearning.com. Headquartered in Stamford, Connecticut, with offices worldwide, Thomson Learning is part of The Thomson Corporation (www.thomson.com), a leading e-information and solutions company in the business, professional, and education marketplaces. The Corporation's common shares are listed on the Toronto and London stock exchanges.

Practice Examination 1 is copyrighted © 1987 by the Department of Personnel, City of New York.
No part of this booklet may be reproduced by any method without the written permission of the Department of Personnel, City of New York.

For more information, contact Peterson's, 2000 Lenox Drive, Lawrenceville, NJ 08648; 800-338-3282; or find us on the World Wide Web at: www.petersons.com/about

Thirteenth Edition

COPYRIGHT © 1999 by Thomson Learning

Previous editions © 1981, 1985, 1990, 1993, 1995, 1997

ALL RIGHTS RESERVED. No part of this work covered by the copyright herein may be reproduced or used in any form or by any means—graphic, electronic, or mechanical, including photocopying, recording, taping, Web distribution, or information storage and retrieval systems—without the prior written permission of the publisher.

For permission to use material from this text or product, contact us by
Phone: 800-730-2214
Fax: 800-730-2215
Web: www.thomsonrights.com

Library of Congress Number available upon request.

ISBN: 0-02-863537-X

Printed in the United States of America

10 9 8 7 6 5 4   02

CONTENTS

PART ONE

The Kind of Work You Will Be Doing 3

Techniques of Study and Test-Taking 9

Firefighter Screening Process 17

    A Typical Notice of Examination 17

    The Application 19

    Medical Standards 25

    Fitness Test and Physical Ability Tests 30

    District of Columbia Fire and Emergency Medical Services Department
        Firefighter Physical Ability Test 37

    Illinois Fire and Police Recruitment Administration, Inc.
        Personnel Aptitude Screening Test 41

    The Interview 45

Glossary of Firefighting Terminology 49

Reading-Based Questions 53

Reading Comprehension for Firefighters 65

Reasoning and Judgment 75

Spatial Orientation 95

Observation and Memory 107

Mechanical Reasoning 127

PART TWO

Practice Examination 1 155

    Answers and Explanations 200

Practice Examination 2 211

    Answers and Explanations 233

Practice Examination 3 243

    Answers and Explanations 265

PART THREE

Practice Examination 4   277
    Answers and Explanations 296

Practice Examination 5   307
    Answers and Explanations 347

Appendix   363
    First Aid 363
    Vehicle Saftety 369

WHAT THIS BOOK WILL DO FOR YOU

ARCO has followed testing trends and methods ever since the firm was founded in 1937. We specialize in books that prepare people for tests. Based on this experience, we have prepared the best possible book to help you score high.

To write this book, we carefully analyzed every detail surrounding firefighter examinations:
- the job itself
- official and unofficial announcements concerning firefighter examinations
- previous examinations, many of which are not available to the public
- related examinations
- technical literature

Can You Prepare Yourself for Your Test?

You want to pass this test. That's why you bought this book. If used correctly, your "self-tutor" will show you what to expect and will give you a speedy brush-up on the subjects tested in your exam. Some of these are subjects not taught in schools at all. Even if your study time is very limited, you should:
- become familiar with the type of examination you will take
- improve your general examination-taking skills
- improve your skills in analyzing and answering questions involving reasoning, judgment, comparison, and evaluation
- improve your speed and skills in reading and in understanding what you read (an important part of your ability to learn and an important part of most tests)

This book will help you prepare by presenting many different types of questions that have appeared on actual firefighter exams.

This book will help you find your weaknesses. Once you know what subjects you're weak in, you can get right to work and concentrate on those areas. This kind of selective study yields maximum test results.

This book will give you the *feel* of the exam. All the questions in the practice exams and in the sections that give instructions in answering reading-based questions, in answering questions of reasoning and judgment, and in dealing with spatial orientation questions were taken from actual previous firefighter exams given at various times in various places. This book will give you experience in answering real questions.

This book will give you confidence as you prepare for the test. It will build your self-confidence as you proceed and will prevent the kind of test anxiety that causes low test scores.

Every city creates its own firefighter exam. The exams are not all alike. The number of questions varies; timing varies; the actual content of the exam varies. Nearly all firefighter exams, however, use the multiple-choice question format and a few basic subject areas such as reading-based questions, reasoning and judgment, use of maps and diagrams, and observation and memory. This book will give you lots of practice with questions in all these areas.

A special feature of ARCO books is an explanation for every answer. You can learn immediately why a correct answer is correct and why the wrong answers are wrong. There's a great deal of information in the answer explanations. Studying answer explanations for all questions, even those you got right, will serve as a course in itself.

PART ONE

CONTENTS

The Kind of Work You Will Be Doing 3
Techniques of Study and Test-Taking 9
Firefighter Screening Process 17
Glossary of Firefighting Terminology 49
Reading-Based Questions 53
Reading Comprehension for Firefighters 65
Reasoning and Judgment 75
Spatial Orientation 95
Observation and Memory 107
Mechanical Reasoning 127

THE KIND OF WORK YOU WILL BE DOING

The Nature of the Work

Every year, fires destroy thousands of lives and property worth millions of dollars. Firefighters help protect the public against this danger. This book is concerned with paid career firefighters; it does not cover the many thousands of volunteer firefighters in communities across the country.

During duty hours, firefighters must be prepared to respond to a fire and to handle any emergency that arises. Because firefighting is dangerous and complicated, it requires organization and teamwork. At every fire, firefighters perform specific duties assigned by a company officer such as a lieutenant, a captain, or another department officer. Firefighters may connect hoselines to hydrants, operate a pump, or position ladders. Because their duties might change several times while the company is in action, they must be skilled in many different firefighting activities such as rescue, ventilation, and salvage. Some firefighters operate fire apparatus, emergency rescue vehicles, and fire-boats. In addition, they help people to safety and administer first aid.

Most fire departments also are responsible for fire-prevention activities. They provide specially trained personnel to inspect public buildings for conditions that might cause a fire. They may check building plans, the number and working condition of fire escapes and fire doors, the storage of flammable materials, and other potential hazards. In addition, firefighters educate the public about fire prevention and safety measures. They frequently speak on this subject before school assemblies and civic groups; in some communities, they inspect private homes for fire hazards.

Between alarms, they have practice drills and classroom training and are responsible for cleaning and maintaining the equipment they use.

Essential Job Tasks Performed by Firefighters

Put on and wear protective equipment

Step off an engine or truck wearing full equipment

Pull uncharged hose off an engine to a hydrant

Advance pre-connected, uncharged $2\frac{1}{2}$-inch hose until it is fully extended

Hold a charged $1\frac{3}{4}$- and/or $1\frac{1}{2}$-inch hose and open the nozzle

While on a ladder, direct water at fire

Remove a K-12 say from truck and start it

Carry a 35-foot extension ladder from truck to the fire scene, with assistance, or carry a 24-foot extension ladder from truck to the fire scene without assistance

Carry a 50-foot section of rolled hose unassisted

Make and unmake coupling connections

Carry a jump bag where needed

Raise a 24-foot and/or 35-foot extension ladder

Extend the fly section of a 24-foot and/or a 35-foot ladder

Operate a charged line from confined spaces

Advance pre-connected, uncharged $1\frac{3}{4}$-inch hose until it is fully extended

Extend the small line to a fire

Operate a fire extinguisher

Remove a smoke ejector (a fan, for example) from the truck; transport and place it in operation

Climb an aerial ladder to a height of 75 to 100 feet wearing full equipment

Run to and from the truck to retrieve tools

Remove and transport electrical extension cords from the truck as needed

- Wrap a hose around hydrant, unscrew the hydrant cap, screw the hose connection to the hydrant, and turn on the hydrant
- Be aware of electrical lines when setting up ladders and directing water streams
- Operate a chain or circular saw during firefighting and rescue operations
- Enter smoking buildings/rooms with a hose in hand while wearing SCBA
- Fight fires in smoky buildings when visibility is poor or nonexistent
- Seek the source of a fire and extinguish
- Prevent the extension of a fire
- Select the correct tool or equipment for various firefighting maneuvers
- Crawl on a floor and if you cannot see, feel for the heat of the fire source
- Safely shut off utility services to buildings in emergency situations
- Determine the stability of supporting surfaces
- Systematically search for trapped persons
- Operate a line from elevated positions
- Climb a fence while wearing full turnout gear and SCBA
- Carry equipment up stairs while wearing full turnout gear and SCBA
- Use equipment (ax, sledge hammer, etc.) to make forcible entries
- Determine when to open roofs, walls, windows and doors
- Use a ceiling hook to pull down a ceiling
- Carry a standpipe rack up numerous flights of stairs while wearing full equipment
- Chop/cut ventilation holes in roof with a fire ax or power saw to let out smoke and heat
- Evacuate persons from a fire area
- Remove persons from entrapments
- Carry or drag a victim out of a building wearing full turnout gear
- Carry or drag victims up or down stairs wearing full turnout gear
- Carry people down ladders wearing full turnout gear
- Remove burned and charred waste
- Re-bed hoses and put them back onto the engine/truck
- Check for smoldering fire inside walls and ceiling after a fire is extinguished
- Lower ground ladders and re-bed them onto the truck
- Extricate people from automobiles
- Respond to hazards related to electrical emergencies

Working Conditions

Firefighters spend much of their time at fire stations, which usually have facilities for dining and sleeping. When an alarm comes in, firefighters must respond rapidly regardless of the weather or hour. They may spend long periods outdoors fighting fires in adverse weather.

Firefighting is among the most hazardous occupations. The job of a firefighter involves the risk of death or injury from sudden cave-ins of floors or toppling walls and danger from exposure to flames and smoke. Firefighters also might come in contact with poisonous, flammable, and explosive gases and chemicals.

In some cities, firefighters are on duty for 24 hours, are then off for 48 hours, and receive an extra day off at intervals. In other cities, they work a day shift of 10 hours for three or four days, a night shift of 14 hours for three or four nights, have three or four days off, and then repeat the cycle. Although in many large cities, particularly in the East, firefighters work a standard 40-hour week, many firefighters average as many as 56 hours per week. In addition to scheduled hours, firefighters often must work extra hours when they are bringing a fire under control. Fire lieutenants and fire captains work the same hours as the firefighters they supervise. Duty hours may include some time when firefighters are free to read, study, or pursue other personal interests.

Places of Employment

More than 280,000 people work as firefighters. Nine out of ten work in municipal fire departments. Some very large cities have several thousand firefighters on the payroll, while many small towns have fewer than 25. Some firefighters work in fire departments on federal installations; others work at airports and in large manufacturing plants.

Training, Other Qualifications, and Advancement

Applicants for municipal firefighting jobs must pass a written test, a medical examination, and tests of strength, physical stamina, and agility, as specified by local regulations. These examinations are open to people who are at least 18 years of age and have a high school education or the equivalent. Those who receive the highest scores on the examinations have the best chances for appointment. Extra credit usually is given for military service. Experience gained as a volunteer firefighter might also improve an applicant's chances for appointment.

As a rule, beginners in large fire departments are trained for several weeks at the department's training center. Through classroom instruction and practical training, the recruits study firefighting techniques, fire prevention, hazardous materials, local building codes, and emergency medical procedures. They also learn how to use axes, saws, chemical extinguishers, ladders, and other firefighting and rescue equipment. After completing this training, they are assigned to a fire company where they are evaluated during a probationary period.

A small but growing number of fire departments have accredited apprenticeship programs that last for three to four years. These programs combine formal, technical instruction with on-the-job training under the supervision of experienced firefighters. Technical instruction covers subjects such as firefighting techniques and equipment, chemical hazards associated with various combustible building materials, emergency medical procedures, and fire prevention and safety.

Many fire departments offer continuing in-service training to members of the regular force. This training may be offered during regular on-duty hours at an easily accessible site connected to the station house alarm system. Such courses help firefighters maintain certain seldom-used skills and permit thorough training on any new equipment the department may acquire.

Most experienced firefighters continue to study to improve their job performance and to prepare for promotion examinations. To progress to higher-level positions, firefighters must acquire expertise in the most advanced firefighting equipment and techniques and in building construction, emergency medical procedures, writing, public speaking, management and budgeting procedures, and labor relations. Fire departments frequently conduct training programs, and some firefighters attend training sessions sponsored by the National Fire Academy on a variety of topics such as executive development, anti-arson techniques, and public fire safety and education. Most states also have extensive firefighter training programs.

Many colleges and universities offer courses such as fire engineering and fire science that are helpful to firefighters, and fire departments often offer firefighters incentives such as tuition reimbursement or higher pay to pursue advanced training. Many fire captains and other supervisory personnel have college training.

Among the personal qualities firefighters need are mental alertness, courage, mechanical aptitude, endurance, and a sense of public service. Initiative and good judgment are extremely important because firefighters often must make quick decisions in emergencies. Because members of a crew eat, sleep, and work closely together under conditions of stress and danger, they should be dependable and be able to get along well with others in a group. Leadership qualities are assets for officers, who must establish and maintain discipline and efficiency as well as direct the activities of firefighters in their companies.

Opportunities for promotion are good in most fire departments. As firefighters gain experience, they can advance to a higher rank. After three to five years of service, they might become eligible for promotion to the grade of lieutenant. The line of further promotion usually is to captain, battalion chief, assistant chief, deputy chief, and finally to chief. Advancement generally depends on scores from a written examination, performance on the job, and seniority. Increasingly, fire departments are using assessment centers—which simulate a variety of actual job performance tasks—to screen for the best candidates for promotion. However, many fire departments require a master's degree—preferably in public administration or a related field—for promotion to positions higher than battalion chief.

Employment Outlook

In the coming years, firefighters are expected to experience increased competition for available jobs. A growing number of people are pursuing firefighting as a career because the educational requirement usually is a high school diploma, earnings are relatively high, and a pension is guaranteed upon retirement. In addition, many people find firefighting to be an exciting, challenging, and rewarding occupation. Consequently, the number of qualified applicants in most areas typically exceeds the number of available jobs. This occurs despite the fact that the written examination and physical requirements eliminate many applicants. The increasing competition for firefighting jobs is expected to persist through the year 2006.

The employment of firefighters is expected to increase more slowly than the average increase for all other occupations. Firefighting jobs are expected to grow at a rate of about 6.1 percent from 1996 to 2006. Compared to the growth rate of about 14 percent for all other jobs, job opportunities for firefighters are likely to be slim. This trend is expected to continue through the year 2006 as fire departments compete with other public service providers for funding. The increased level of specialized training required in this occupation has made it increasingly difficult for volunteer firefighters to remain qualified. As a result, an expected increase in the number of paid firefighter positions versus volunteer firefighters will result in the most growth. Large, urban fire departments are expected to experience the least amount of growth.

In recent years, firefighters have gotten involved in much more than preventing fires. In response to the expanding role of firefighters, some municipalities have combined fire prevention, public fire education, safety, and emergency medical services into a single organization. Some local and regional fire departments are being consolidated into county-wide public safety establishments. This has allowed for reduction of overhead, reduction of administrative staff, and establishment of consistent training standards and work procedures.

Turnover of firefighter jobs is particularly low at this time. This is somewhat unusual considering that the job of a firefighter is hazardous and requires a relatively limited investment in formal education. Most job openings are expected to arise from the need to replace firefighters who retire, who stop working for other reasons, or who transfer to other occupations.

Layoffs of firefighters continue to be uncommon. Fire protection is an essential service for all citizens. As a result, city officials are likely to feel considerable pressure to expand or at least preserve the level of fire-protection coverage. When budget cuts do occur, local fire departments usually cut expenses by postponing equipment purchases or by not hiring new firefighters rather than by laying off staff.

Earnings

Many career firefighters and company officers are unionized and belong to the International Association of Firefighters. Many chief officers belong to the International Association of Fire Chiefs. These unions have been highly effective in securing wages and benefits for their members to compensate the efforts they make, the responsibilities they take on, and the dangers they face.

In 1996, the average weekly earnings for firefighting occupations were around $658. The middle 50 percent earned between $513 and $832 weekly. The lowest 10 percent earned less than $387, while the highest 10 percent earned more than $979. The average annual salary for all firefighters in the federal government in nonsupervisory, supervisory, and managerial positions was about $28,800 in 1996. Fire lieutenants and fire captains may earn considerably more.

Firefighters who average 53 or more hours per week during their work period, which typically ranges from 7 to 28 days, are required to be paid for overtime. Firefighters often earn overtime for working extra shifts to maintain minimum staffing levels or for special emergencies.

Firefighters receive benefits that usually include medical and liability insurance, vacation and sick leave, and some paid holidays. Almost all fire departments provide protective clothing (helmets, boots, and coats) and breathing apparatus. Many departments also provide dress uniforms. Firefighters generally are covered by pension plans that often provide retirement at half pay at age 50, after 25 years of service, or if disabled in the line of duty.

Related Occupations

Firefighters work to prevent fires and to save lives and property when fires and other emergencies such as explosions and chemical spills occur. Related fire-protection occupations include wild-land firefighters and fire-protection engineers, who identify fire hazards in homes and workplaces and design prevention programs and automatic fire detection and extinguishing systems. Other occupations in which workers respond to emergencies include police officers and emergency medical technicians.

Sources of Additional Information

Information about obtaining a job as a firefighter is available from local civil service offices or fire departments.

Information about a career as a firefighter can be obtained from:

International Association of Fire Chiefs, 1329 18th St. NW, Washington, DC 20036.

International Association of Fire Fighters, 1750 New York Ave. NW, Washington, DC 20006.

Information about firefighter professional qualifications can be obtained from:

National Fire Protection Association, Batterymarch Park, Quincy, MA 02269.

Additional information about the salaries and work hours of firefighters in various cities is published annually by the International City Management Association in its *Municipal Yearbook*, which is available in many libraries.

TECHNIQUES OF STUDY AND TEST-TAKING

You want to become a firefighter. That's great. Every city, every town, and every village needs qualified and enthusiastic new recruits to maintain staffing at a full level. You have made a good start toward joining the firefighting force by buying this book.

This book has been carefully researched and written to help you through the qualifying process. The information in this book will prepare you for the written exam and will give you valuable tips towards preparing yourself for the physical performance exam.

It is important for you to allow the book to help you. Read it carefully. Do not skip over any information. You must know what to expect and then prepare yourself for it. If you are prepared, you can feel self-confident. If you feel confident, you can answer questions quickly and decisively, finish the exam, and earn a high score. If you feel confident, you can enter the physical performance course without hesitation and can prove that you are fit for the job.

Preparing for the Exam

1. **Make a study schedule.** Assign yourself a period of time each day to devote to preparation for your firefighter exam. A regular time is best, but the important thing is daily study.
2. **Study alone.** You will concentrate better when you work by yourself. Keep a list of questions you find puzzling and points of which you are unsure to talk over with a friend preparing for the same exam. Plan to exchange ideas at a joint review session just before the test.
3. **Eliminate distractions.** Choose a quiet, well-lit spot as far away as possible from the telephone, television, and family activities. Try to arrange not to be interrupted.
4. **Begin at the beginning.** Read. Underline points that you consider to be significant. Make margin notes. Flag the pages you think are especially important with little Post-it™ Notes. We have highlighted some pages with colored paper. Read these pages more than once.
5. **Concentrate on the information and instruction chapters.** Read the Glossary of Firefighting Terminology on page 51. Learn the vocabulary of the job. Get yourself psyched for the whole world of firefighting. Learn how to handle reading-based questions. Focus on the approach to eliminate wrong answers. This information is important for answering all multiple-choice questions, but it is especially vital to questions of reasoning and judgment.
6. **Answer the practice questions in each chapter.** Study the answer explanations. You can learn a great deal from answer explanations. Even when you have answered correctly, the explanation might bring out points that had not occurred to you. This same suggestion—read *all* the explanations—applies throughout this book to the exams as well as the instructional chapters.
7. **Try the practice exams.** When you believe you are well prepared, move on to the exams. If possible, answer an entire exam in one sitting. If you must divide your time, divide it into no more than two sessions per exam.

When you take the practice exams, treat them with respect. Consider each as a dress rehearsal for the real thing. Time yourself accurately and do not peek at the correct answers. Remember, you are taking these for practice. They will not be scored and do not count, so learn from them. Learn to think; learn to reason like an effective firefighter; learn to pace yourself so you can answer all the questions. Then learn from the explanations.

IMPORTANT: Do not memorize questions and answers. Any question that has been released will not be used again. You might run into questions that are very similar, but you will not be tested with any of these exact questions. These questions will provide good practice, but they will not give you the answers to any of the questions on your exam.

How to Take an Exam

1. **Get to the examination room about 10 minutes ahead of time.** You'll get a better start when you are accustomed to the room. If the room is too cold, too warm, or not well ventilated, call these conditions to the attention of the person in charge.
2. **Make sure you read the instructions carefully.** In many cases, test-takers lose points because they misread an important part of the directions. (An example would be selecting the *incorrect* choice instead of the *correct* choice.)
3. **Don't be afraid to guess.** The best policy, of course, is to pace yourself so you can read and consider each question. Sometimes this does not work. Most civil service exam scores are based on the number of questions answered correctly. This means that a wild guess is better than a blank space. There is no penalty for a wrong answer, and you just might guess right. If you see that time is about to run out, mark all the remaining spaces with the same answer. According to the law of averages, some will be right.

    You have bought this book, however, with entire exams for practice answering firefighter questions. Part of your preparation is learning to pace yourself so you need not answer randomly at the end. Far better than a wild guess is an educated guess. You make this kind of guess not when you are pressed for time but when you are not sure of the correct answer. Usually, one or two of the choices are obviously wrong. Eliminate the obviously wrong answers and try to reason among those remaining. Then, if necessary, guess from the smaller field. The odds of choosing a right answer increase if you guess from a field of two rather than from a field of four. When you make an educated guess or a wild guess in the course of the exam, you might want to make a note next to the question number in the test booklet. If there is time, you can go back for a second look.
4. **Reason your way through multiple-choice questions very carefully and methodically.** Here are a few samples we can "walk through" together:

    I. On the job, your supervisor gives you a hurried set of directions. As you start your assigned task, you realize you are not quite clear on the directions given to you. The best action to take would be to
        (A) continue with your work, hoping to remember the directions and to do the best you can
        (B) ask a co-worker in a similar position what he or she would do
        (C) ask your supervisor to repeat or clarify certain directions
        (D) go on to another assignment

In this question, you are given four possible answers to the problem described. Though the four choices are all *possible* actions, it is up to you to choose the *best* course of action in this particular situation.

Choice (A) will likely lead to a poor result. Given that you do not recall or understand the directions, you would not be able to perform the assigned task properly. Keep choice (A) in abeyance until you have examined the other alternatives. It could be the best of the four choices given.

Choice (B) also is a possible course of action, but is it the best? Consider that the co-worker you consult has not heard the directions. How could he or she know? Perhaps his or her degree of incompetence is greater than yours in this area. Of choice (A) and choice (B), the better of the two is still choice (A).

Choice (C) is an acceptable course of action. Your supervisor will welcome your questions and will not lose respect for you. At this point, you should hold (C) as the best answer and eliminate (A).

The course of action in (D) is decidedly incorrect because the job at hand would not be completed. Going on to something else does not clear up the problem; it simply postpones having to make a necessary decision.

After careful consideration of all choices given, choice (C) stands out as the best possible choice of action. You should select (C) as your answer.

Every question is written about a fact or an accepted concept. This question indicates the concept that, in general, most supervisory personnel appreciate subordinates questioning directions that might not have been fully understood. This type of clarification precludes subsequent errors on the part of the subordinates. On the other hand, many subordinates are reluctant to ask questions for fear that their lack of understanding will detract from their supervisor's evaluation of their abilities.

The supervisor, therefore, has the responsibility to issue orders and directions in such a way that subordinates will not be discouraged from asking questions. This is the concept on which the sample question was based.

If you are familiar with this concept, you should have no trouble answering the question. If you are not familiar with it, however, the method outlined here of eliminating incorrect choices and selecting the correct one should prove successful for you.

You have now seen how important it is to identify the concept and the key phrase of the question. Equally important, or perhaps even more so, is identifying and analyzing the key word—the qualifying word—in a question. This word is usually an adjective or adverb. Some of the most common key words are *most*, *least*, *best*, *highest*, *lowest*, *always*, *never*, *sometimes*, *most likely*, *greatest*, *smallest*, *tallest*, *average*, *easiest*, *most nearly*, *maximum*, *minimum*, *only*, *chiefly*, *mainly*, *but*, and *or*. Identifying these key words usually is half the battle in understanding and, consequently, answering all types of exam questions.

Let's use the elimination method on some additional questions:

II. On the first day you report for work after being appointed as a firefighter, you are assigned to routine duties that seem to you to be very petty in scope. You should
  (A) perform your assignment perfunctorily while conserving your energies for more important work in the future
  (B) explain to your superior that you are capable of greater responsibility
  (C) consider these duties an opportunity to become thoroughly familiar with the firehouse
  (D) try to get someone to take care of your assignment until you have become thoroughly acquainted with your new associates

Once again, you are confronted with four possible answers from which you are to select the *best* one.

Choice (A) will not lead to getting your assigned work done in the best possible manner in the shortest possible time. This would be your responsibility as a newly appointed firefighter, and the likelihood of getting to do more important work in the future following the approach stated in this choice is remote. Since this is an (A) choice, however, you must hold it aside because it might turn out to be the best of the four choices given.

Choice (B) is better than choice (A) because your superior might not be familiar with your capabilities at this point. You therefore should drop choice (A) and retain choice (B) because, once again, it might be the best of the four choices.

The question clearly states that you are newly appointed. Therefore, wouldn't it be wise to perform whatever duties you are assigned in the best possible manner? This way, you not only use the opportunity to become acquainted with firehouse procedures but also to demonstrate your abilities. Choice (C) contains a course of action that will benefit you and the firehouse in which you are working because it gets needed work done. At this point, you should drop choice (B) and retain choice (C) because it is by far the better of the two.

The course of action in choice (D) is not likely to get the assignment completed, and it will not enhance your image to your fellow firefighters. Choice (C), when compared to choice (D), is far better and should be selected as the *best* choice.

Now let's examine a question that appeared on a police officer examination:

III. An off-duty police officer in civilian clothes riding in the rear of a bus notices two teenage boys tampering with the rear emergency door. The most appropriate action for the officer to take is to
   (A) tell the boys to discontinue their tampering, pointing out the dangers to life that their actions might create
   (B) report the boys' actions to the bus operator and let the bus operator take whatever action is deemed best
   (C) signal the bus operator to stop, show the boys his officer's badge, and then order them off the bus
   (D) show the boys his officer's badge, order them to stop their actions, and take down their names and addresses

Before considering the answers to this question, you must accept the well-known fact that a police officer is always on duty to uphold the law even though he or she might be technically off duty.

In choice (A), the course of action taken by the police officer will probably serve to educate the boys and will get them to stop their unlawful activity. Because this is only the first choice, you will hold it aside.

In choice (B), you must realize that the authority of the bus operator in this instance is limited. He can ask the boys to stop tampering with the door but that is all. The police officer can go beyond that point. Therefore, you should drop choice (B) and continue to hold choice (A).

Choice (C), as a course of action, will not have a lasting effect. What is to stop the boys from boarding the next bus and continuing their unlawful action? Therefore, drop choice (C) and continue to hold choice (A).

Choice (D) might have some beneficial effects, but it would not deter the boys from continuing their actions in the future. When you compare choice (A) with choice (D), (A) is the better one overall and should therefore be your selection.

The next example illustrates a type of question that has gained popularity in recent examinations and that requires a two-step evaluation. First, you must evaluate the condition in the question as being "desirable" or "undesirable." After the determination has been made, you then are left with making a selection from two choices instead of the usual four.

IV. A visitor to an office in a city agency tells one of the office aides that he has an appointment with the supervisor of the office who is expected shortly. The visitor asks for permission to wait in the supervisor's private office, which is unoccupied at the moment. For the office aide to allow the visitor to do so would be
   (A) desirable; the visitor would be less likely to disturb the other employees or to be disturbed by them
   (B) undesirable; it is not courteous to permit a visitor to be left alone in an office
   (C) desirable; the supervisor might want to speak to the visitor in private
   (D) undesirable; the supervisor might have left confidential papers on the desk

You first must evaluate the course of action on the part of the office aide. Permitting the visitor to wait in the supervisor's office is *very* undesirable. There is nothing said of the nature of the visit; it might be for a purpose that is not friendly or congenial. There might be papers on the supervisor's desk that he or she does not want the visitor to see or to have knowledge of. Therefore, at this point, you have to make a choice between (B) and (D).

This is definitely not a question of courtesy. Although all visitors should be treated with courtesy, permitting the visitor to wait in the supervisor's office is not the only possible act of courtesy. Another comfortable place could be found for the visitor to wait.

Choice (D) contains the exact reason for evaluating this course of action as being undesirable. When compared with choice (B), choice (D) is far better.

One word of caution: After you have made a determination that a course of action is desirable or undesirable, confine your thoughts to the situations illustrating your choice. Do not revert to again considering the other extreme or you will flounder and will surely not select the correct answer.

4. **Mark the answer sheet clearly**. Unless you have applied to a fire force in a very small locality, your exam will be machine scored. You will mark your answers on a separate answer sheet. Each exam in this book has its own answer sheet so you can practice marking all your answers in the correct way. Tear out the answer sheet before you begin each exam. Do not try to flip pages back and forth. Because your answer sheet will be machine scored, you must fill it out clearly and correctly. You cannot give any explanations to the machine. This means:
   - You must blacken your answer space firmly and completely. ● is the only correct way to mark the answer sheet. ◐, ⊗, ⊘, and ⊘ are all unacceptable. The machine might not read them at all.
   - You must mark only one answer for each question. If you mark more than one answer, you will be considered wrong even if one of the answers is correct. (See note on page 12.)
   - If you change your mind, you must erase your mark. Attempting to cross out an incorrect answer like this ● will not work. You must erase any incorrect answer completely. An incomplete erasure might be read as a second answer.
   - All of your answers should be in the form of blackened spaces. The machine cannot read English. Do not write any notes in the margins. If you have done any figuring in the margins of the test booklet itself, be sure to mark the letter of your answer on the answer sheet. Correct answers in the test booklet do not count. Only the answer sheet is scored.
   - MOST IMPORTANT: Answer each question in the correct place. Question 1 must be answered in space 1, question 52 in space 52. If you skip an answer space and mark a series of answers in the wrong places, you must erase all the answers and do the questions over, marking your answers in the proper places. You cannot afford to use the limited time in this way. Therefore, as you answer *each* question, look at its number and make sure you are marking your answer in the space with the same number. The risk involved in slipping out of line on the answer sheet is the reason we recommend that you answer every question in order, even if you have to guess. Make an educated guess if you can. If not, make a wild guess. Just mark the question number in the test booklet so you can try again if there is time.

TEST DAY STRATEGY

On the examination day assigned to you, allow the test itself to be the main attraction of the day. Do not squeeze it in between other activities. Arrive rested, relaxed, and on time. In fact, plan to arrive a little bit early. Leave plenty of time for traffic tie-ups or other complications that might upset you and interfere with your test performance.

In the test room, the examiner will provide forms for you to fill out. He or she will give you the instructions you must follow in taking the examination. The examiner will tell you how to fill in the grids on the forms. Time limits and timing signals will be explained. If you do not understand any of the examiner's instructions, ASK QUESTIONS. Make sure that you know exactly what to do.

During the examination, you must follow instructions exactly. Fill in the grids on the forms carefully and accurately. Filling in the wrong grid might lead to a loss of veterans' credits to which you might be entitled or to an incorrect address for your test results. Do not begin until you are told to do so. Stop as soon as the examiner tells you to stop. Do not turn pages until you are told to do so. Do not go back to parts you have already completed. Any infraction of the rules is considered cheating. If you cheat, your test paper will not be scored, and you will not be eligible for appointment.

Once the signal has been given and you begin the exam, READ every word of every question. Be alert for exclusionary words that might affect your answer—words like "not," "most," "least," "all," "every," and "except."

READ all the choices before you mark your answer. It is statistically true that most errors are made when the last choice is the correct answer. Too many people mark the first answer that seems correct without reading through all the choices to find out which answer is *best*.

The following list summarizes the suggestions we already have given for taking this exam. Read the suggestions now and again before you attempt the exams in this book. Review them before you take the actual exam. You will find them all useful.

1. Mark your answers by completely blackening the answer space of your choice.
2. Mark only ONE answer for each question, even if you think more than one answer is correct. You must choose only one. If you mark more than one answer, the scoring machine will consider you wrong even if one of your answers is correct.
3. If you change your mind, erase your first response completely. Leave no doubt as to which answer you have chosen.
4. If you do any figuring in the test booklet or on scratch paper, make sure to still mark your answer on the answer sheet.
5. Check often to make sure that the question number matches the answer space number and that you have not skipped a space by mistake. If you do skip a space, you must erase all the answers after the skip and answer all the questions again in the right places.
6. Answer every question in order but do not spend too much time on any one question. If a question seems to be "impossible," do not take it as a personal challenge. Guess and move on. Remember that your task is to answer correctly as many questions as possible. You must apportion your time to give yourself a fair chance to read and answer all the questions. If you guess at an answer, mark the question in the test booklet so you can find it easily if time permits.
7. Guess intelligently if you can. If you do not know the answer to a question, eliminate the answers you know are wrong and guess from the remaining choices. If you have no idea whatsoever as to the answer to a question, guess anyway. Choose an answer other than the first. The first choice generally is the correct answer less often than the other choices. If your answer is a guess, either an educated guess or a wild one, mark the question in the question booklet so you can give it a second try if time permits. If time is about to run out and you have not had a chance to answer all the questions, mark all the remaining questions with the same answer. According to the law of averages, you should get 25 percent of them right.
8. If you happen to finish before time is up, check to make sure that each question is answered in the right space and that there is only one answer for each question. Return to the difficult questions you marked

in the booklet and try them again. There is no bonus for finishing early so use all your time to perfect your exam paper.

***Note:** A recent administration of the Firefighter Exam in New York City experimented with a new method of marking answers. Instead of marking only one answer for each question, as is done in conventional multiple-choice exams, test-takers were told to mark first choice, second choice, and third choice for each question. Partial credit was then awarded to questions for which a test taker marked the correct answer as a second or third choice. This change in the marking system was reported in the newspapers only days before the exam and caught everyone by surprise. Detailed instructions for marking the answer sheet accompanied the exam and answer sheet. Even so, many applicants had difficulty with the complicated system. Test makers do experiment. The lesson to you is to READ CAREFULLY. Do not assume that all rules and procedures will remain as they always have been. Read all instructions. Ask questions before the exam begins. Be certain that you understand exactly what you are to do. Then follow through as directed.

FIREFIGHTER SCREENING PROCESS

A Typical Notice of Examination

The general provisions of the notice of examination and the general examination regulations of the New York City Department of Personnel are part of this notice of examination. They are posted, and copies are available, at the Application Section of the Department of Personnel at 18 Washington Street, New York, NY 10004.

This position is open to both men and women.

Candidates appointed to this position may be required to work rotating tours or shifts including Saturdays, Sundays, and holidays.

Promotion Opportunities: Employees in the title of Firefighter are accorded promotion opportunities, when eligible, to the title of Lieutenant, Fire Department.

REQUIREMENTS

MINIMUM REQUIREMENTS: By the date of appointment, you must be a United States citizen and possess a four-year high school diploma or its equivalent.

LICENSE REQUIREMENT: On the date of appointment as a firefighter, possession of a valid New York State Driver License is required.

AGE REQUIREMENT: Pursuant to Section 15-103 of the Administrative Code, only persons who have passed their 18th birthday but not their 29th birthday on the date of filing qualify for this examination. However, only persons who have reached their 21st birthday shall be appointed.

In accordance with Section 54 of the New York Civil Service Law, no applicant within six months of the minimum age for filing can be prohibited from taking the written test. In summary, the age requirement means that you must not have reached your 29th birthday by September 2 but you must be at least $17\frac{1}{2}$ years of age by September 22 to take the December 12 exam.

EXCEPTION TO THE AGE REQUIREMENT: All persons who were engaged in military duty, as defined in Section 243 of the Military Law, may deduct from their actual age the length of time spent in such military duty, provided the total deduction for military duty does not exceed six years, of which not more than four years may be for voluntary service after May 7, 1975.

INVESTIGATION: At the time of appointment and at the time of investigation, candidates must present all the official documents and proof required to qualify. At the time of investigation, candidates will be required to present proof of birth by transcript of records from the Bureau of Vital Statistics or other satisfactory evidence. Proof of military service and similar documents may be required. Failure to present required documents, including proof of educational or license requirements, will result in disqualification for appointment. Any willful misstatement or omission will be cause for disqualification.

CHARACTER AND BACKGROUND: Proof of good character and satisfactory background will be an absolute prerequisite to appointment. In accordance with the provisions of law, persons convicted of a felony are not eligible for positions in the uniformed force of the fire department. In addition, the rules of the Department of Personnel provide that no person who has been dishonorably discharged from the Armed Forces shall be certified for appointment as a firefighter.

MEDICAL TEST: Candidates will be required to pass a medical test prior to appointment.

RESIDENCY REQUIREMENT: The New York Public Officers Law requires that any person employed as a firefighter in the New York City Fire Department be a resident of the City of New York or a resident of Nassau, Westchester, Suffolk, Orange, Rockland, or Putnam counties.

PROOF OF IDENTITY: Under the Immigration Reform and Control Act of 1986, you must be able to prove your identity and your right to obtain employment in the United States prior to employment with the City of New York.

JOB DESCRIPTION

Duties and Responsibilities: Under supervision, assists in the control and extinguishment of fires and in the enforcement of laws, ordinances, rules, and regulations regarding the prevention, control, and extinguishment of fires; performs inspection, investigation, and regulative duties connected with fire prevention, control, and extinguishment of fires; performs all additional functions for the rank prescribed by orders, directives, or regulations of the fire department and performs special duties or assignments as directed by the Fire Commissioner.

TEST INFORMATION

Tests: There will be a written test of the multiple-choice type and a physical test. Only candidates who achieve the passing mark of 70 percent on the written test will be summoned for the physical test. The weight of the written test will be announced by the date of the written test.

The written test will be of the multiple-choice type and will be designed to test the candidates' ability to read and understand English and his or her capacity to learn and perform the work of a firefighter. It might include questions about understanding job information; applying laws, rules, and regulations to job situations; recognizing appropriate behavior in job situations; understanding mechanical devices; and remembering details of a floor layout.

The physical test will be designed to test the candidates' capacity to perform the physical aspects of the work of a firefighter.

Medical and Physical Standards: Eligible persons must meet the medical and physical standards to be established and posted on the Department of Personnel Bulletin Board. Candidates must pass the physical test and a qualifying medical test. Eligible persons will be rejected for any physical, mental, or medical condition that will prevent them from performing in a reasonable manner the duties of a firefighter.

The Application

The application form you must file could be as simple as the New York City application form reproduced on the next page or could be considerably more complex and more detailed than the Experience Paper that follows it.

What is important is that you fill out your application form neatly, accurately, completely, and truthfully.

Obviously, neatness counts. If your application is misread, it is likely to be misfiled or misinterpreted to your detriment. If your application is extremely sloppy, it brands you as a slob. No one wants to hire a slob.

Accuracy is important. A single wrong digit in your social security number or exam number can misroute your application forever. If your inaccuracy is discovered and corrected, you nevertheless carry with you the label of "careless."

Completeness is a requirement. An incomplete application gets tossed into the wastebasket. No one will take the time to chase you down to fill in the blanks.

The need to tell the truth should be self-evident. Certainly, you should describe your duties and responsibilities in the best possible light but do not exaggerate. Do not list a degree you did not receive. Do not give a graduation date if you did not graduate. If your application form asks for reasons for leaving previous employment, give a very brief but truthful reason. Any statement you make on an application form is subject to verification. There are staff members at the Civil Service office whose job it is to check with schools and former employers. These investigators ask questions about your job performance, attendance record, duties, and responsibilities.

Any misstatements that are not picked up by background investigators are likely to be unmasked during an interview. The last step of the screening process is likely to be a personal interview before one person or before a panel of interviewers. The interviewer(s) will sit with a copy of your application form and will ask you some questions based upon it. This will be your opportunity to explain unfavorable items such as dismissal from a job or frequency of job changes. If you have made false claims or have stretched the truth too far, this is the point at which you are likely to trip yourself up. It is difficult to maintain a lie under the pressure of pointed questioning. The interviewers are prepared and are at ease. You are in an unfamiliar situation and are nervous about qualifying for a job. Do not make it harder for yourself. Be open and truthful on the application; defend yourself at the interview.

APPLICATION FOR OPEN COMPETITIVE EXAMINATION
(See Instructions on Reverse Side)

CITY OF NEW YORK
DEPARTMENT OF PERSONNEL
APPLICATION SECTION

EXAM NO.

YOUR SOCIAL SECURITY NO. (check your card before entering your number)

EXAM TITLE

FEE (See Item 3 on reverse)
$

YOUR FIRST NAME | Middle Initial | YOUR LAST NAME

YOUR STREET ADDRESS (INCLUDE APT. NO., BLDG NO., OR CARE OF)

CITY OR TOWN | STATE | ZIP CODE

IF YOU LIVE IN N.Y.C., CHECK THE BOROUGH THAT YOU LIVE IN.

MANHATTAN — M
BRONX — X
BROOKLYN — K
QUEENS — Q
STATEN ISLAND — R

CHECK ALL BOXES THAT APPLY TO YOU.

☐ I CLAIM VETERANS' CREDIT. (SEE ITEM 7 ON REVERSE.) — V
☐ I CLAIM DISABLED VETERANS' CREDIT. (MINIMUM 10% DISABILITY) (SEE ITEM 7 ON REVERSE.) — D
☐ I AM A SABBATH OBSERVER. (SEE ITEM 4 ON REVERSE.) — S
☐ I HAVE A HANDICAP REQUIRING SPECIAL TESTING ACCOMMODATIONS. (SEE ITEM 5 ON REVERSE.) — H

DECLARATION: I declare that the statements on this form are true. I further declare that if I have made a claim for Veterans' or Disabled Veterans' credit that I meet the requirements for such credits as described on the reverse side. I am aware that I must prove that I am entitled to the credit I claim or my employment may be terminated.

Your Signature _____

REMINDER

- DID YOU ENCLOSE THE CORRECT FEE?
- DID YOU FILL OUT COMPLETELY AND ATTACH ALL OTHER FORMS REQUIRED BY NOTICE OF EXAMINATION?
- IS YOUR SOCIAL SECURITY NUMBER CORRECT?
- DID YOU SIGN YOUR NAME?
- DID YOU ANSWER EVERY QUESTION?
- DID YOU GIVE YOUR COMPLETE ADDRESS?
- DID YOU INCLUDE YOUR CORRECT ZIP CODE?

AN INCOMPLETE APPLICATION MAY NOT BE PROCESSED. BE SURE TO INDICATE YOUR FULL NAME, SOCIAL SECURITY NUMBER, ADDRESS AND ZIP CODE.

BE SURE TO READ THE INSTRUCTIONS ON THE REVERSE SIDE.

DP-848C (R 1/86)

FIREFIGHTER SCREENING PROCESS 21

INSTRUCTIONS FOR FILLING OUT THIS APPLICATION FORM

Be sure to read the Notice of Examination carefully before completing this form.

1. **FORMS** — If the Notice of Examination calls for Experience Form A and/or other forms, these forms must be included with your application, or your application cannot be accepted.

2. **ADDRESS** — Give your full mailing address, including building, apartment number, or "in care of" information where needed. (If you change your address after applying, write to the Examining Service Division of the Department of Personnel, Room 216, 220 Church Street, New York, N.Y. 10013, with your new address, plus your Social Security Number and the number and title of the examination for which you applied.)

3. **FEE** — Enclose the correct fee made out to the N.Y.C. Department of Personnel. The amount of the fee is stated in the Notice of Examination for this examination. DO NOT MAIL CASH. If paying by check: write your name, social security number, home address, and the examination number on the front of the check. A fee is not charged if you are a N.Y.C. resident receiving public assistance from the N.Y.C. Department of Social Services. To have the fee waived you must present or enclose a clear photocopy of your current Medicaid Card. The photocopy must accompany the application, even if it is filed in person.

4. **SABBATH OBSERVER** — If because of religious belief you cannot take a test on the scheduled date, you must come or write to: Examining Service Division (Room 216): N.Y.C. Department of Personnel, 220 Church Street, New York, N.Y. 10013 no later than five work days prior to the test date to request an alternate date. All requests must be accompanied by a signed statement on letterhead from your religious leader certifying to your religious observance.

5. **HANDICAP** — If you have a handicap which will interfere with your ability to take this test without special accommodations, amanuensis, or other assistance, you must submit a written request for specific special accommodations to the Examining Service Division (Room 216); N.Y.C. Department of Personnel, 220 Church Street, New York, N.Y. 10013. The request must be received no later than 15 work days prior to the test date. A physician or agency authorized for this purpose must corroborate the specific nature of your handicap and must justify the need for the special accommodations you request. For more details, consult Regulation E.8 of the General Examination Regulations, which is available from the Application Section of the Department of Personnel.

6. **MAIL** — Check the Notice of Examination to see if applications are accepted by mail. If you mail your application, use a 4¼" x 9½" (legal size) envelope, and address to:

    N.Y.C. Department of Personnel
    Application Section

    Your application MUST BE RECEIVED IN THIS DEPARTMENT BY THE LAST DAY OF FILING.

7. **VETERANS' CREDIT** — FOR VETERANS' OR DISABLED VETERANS' CREDIT YOU MUST:

    a. Have served on active duty, other than for training purpose, in the Armed Forces of the United States during:
       - December 7, 1941 - September 2, 1945; or
       - June 26, 1950 - January 31, 1955; or
       - January 1, 1963 - May 7, 1975; and
    b. Be a resident of New York State at the time of list establishment; and
    c. Have an honorable discharge or have been released under honorable conditions.
    d. For Disabled Veterans' Credit, in addition to a, b, and c above, at the time the examination list is established you must receive or be entitled to receive at least 10% compensation from the V.A. for a disability incurred in time of war. The V.A. must also certify that the disability is permanent. If it is not permanent, you must have been examined by the V.A. within one year of the establishment of the examination list.

    NOTES

    1. You may use Veterans' or Disabled Veterans' Credit only once after January 1, 1951 for appointment or promotion from a City, State or County civil service list.
    2. The above is only a summary of necessary conditions. The complete provisions are contained in statutory and/or decisional law.

SOCIAL SECURITY NO. ___/___/___	DEPARTMENT OF PERSONNEL **EXPERIENCE PAPER** _____ Exact Title of Examination	FORM A EXAMINATION NO. _____

DO NOT WRITE IN THIS SPACE

Not qualified under terms of advertisement	QUALIFIED: Admitted for further examination pending final determination of qualifications.	RATING

To Investigator: Verify items checked and also data mentioned here ....

☐ If this box is checked, submit a report with full details to ....

FOLLOW THESE INSTRUCTIONS CAREFULLY:

a. Type, print or write (in blue or black ink) the information requested.
b. Be COMPLETE, CONCISE AND ACCURATE. The information you enter below will be used to determine your rating on training and experience. If statements of material facts are found to be false, exaggerated or misleading, you may be disqualified.
c. DO NOT REVEAL YOUR IDENTITY in any way on this experience paper. If you are or have been in business for yourself or have been employed by a relative of the same name, write "self" or the relationship of the relative in the space headed "Name of Employer".

1. **LICENSE OR REGISTRATION:** If a license or registration is required, answer the following.
   a. Title of lic. or reg. you possess which is valid in this State .... License or Registration No. ....
   b. Name of issuing agency ....
   c. Date of original issue .... Date last renewed or registered .... Renewal number .... Date of expiration ....

2. **EDUCATION:**   3. **DATE OF BIRTH*** ....

Give Name and Locations of Schools Below	Day or Night	From Mo. Yr.	To Mo. Yr.	Were you Graduated (Yes or No)	Degree Received	Total Credits Completed	Major Subject	No. of Credits in Major
High School or Trade School								
College or other School								

4. List below any courses which you have passed which: (a) are necessary to meet the minimum requirements for the position; (b) provide training appropriate to the position; (c) are in related fields.

NAME OF COLLEGE OR INSTITUTION	Course Number	EXACT TITLE OF COURSE (Place "G" after graduate courses)	No. of Credits	Date Completed

*The Human Rights Law prohibits discrimination on the basis of age, creed, color, national origin, sex, disability or marital status of any individual. The law allows certain age or sex specifications if based upon a bona fide occupational qualification or statutory authorization.

QUALIFYING EMPLOYMENT. List in chronological order those positions you have held which tend to qualify you for the position sought. Begin with the present or most recent position. Use a separate block for each position. List as a separate employment every material change of duties although with the same employer. Include pertinent experience or training in the armed forces.

1	Dates of Employment (give month and year) From To	Length of Employment Yrs. Mos.	Exact Title of Your Position	Starting Salary per Last Salary per
	Name and Address of Employer		Nature of Employer's Business	
			Title (not name) of your Immediate Superior	
	Number and Titles of Employees You Supervise		If with the City or State was this a provisional appointment? Yes ☐ No ☐	No. of Hours Worked per Week
	Describe Your Duties			

2	Dates of Employment (give month and year) From To	Length of Employment Yrs. Mos.	Exact Title of Your Position	Starting Salary per Last Salary per
	Name and Address of Employer		Nature of Employer's Business	
			Title (not name) of your Immediate Superior	
	Number and Titles of Employees You Supervise		If with the City or State was this a provisional appointment? Yes ☐ No ☐	No. of Hours Worked per Week
	Describe Your Duties			

3	Dates of Employment (give month and year) From To	Length of Employment Yrs. Mos.	Exact Title of Your Position	Starting Salary per Last Salary per
	Name and Address of Employer		Nature of Employer's Business	
			Title (not name) of your Immediate Superior	
	Number and Titles of Employees You Supervise		If with the City or State was this a provisional appointment? Yes ☐ No ☐	No. of Hours Worked per Week
	Describe Your Duties			

4	Dates of Employment (give month and year) From To	Length of Employment Yrs. Mos.	Exact Title of Your Position	Starting Salary per Last Salary per
	Name and Address of Employer		Nature of Employer's Business	
			Title (not name) of your Immediate Superior	
	Number and Titles of Employees You Supervise		If with the City or State was this a provisional appointment? Yes ☐ No ☐	No. of Hours Worked per Week
	Describe Your Duties			

OTHER EXPERIENCE:	List here only the other employments you have had which were not included before.				
Dates of Employment		Name and Address of Employer	Business of Employer	Title of Your Position and Highest Salary	Brief Description of Duties You Performed
From	To				

Any additional qualifications you have for the position may be listed here; for example: field work in connection with training, authorship or co-authorship of books or of articles in recognized journals, special certification from professional boards, other professional attainments, etc. (Caution: Do not mention your name.) Additional courses or experience may also be listed here.

IF YOU HAVE ADDITIONAL QUALIFYING EXPERIENCE, ATTACH SEPARATE PAPER USING THE SAME FORMAT

Medical Standards

Firefighting is a physically stressful occupation. Firefighters who spend much of their time doing relatively undemanding physical tasks in the firehouse and who spend a good deal of the time in sedentary activity such as card playing and watching television are suddenly called to put forth major physical exertion. The rapid and extreme shifts in demands on the body of a firefighter result in real punishment. This is why the selection process places such emphasis on medical condition and on physical fitness and capacity.

The actual firefighting activity involves heavy lifting and carrying while encumbered by heavy protective gear under conditions of extreme temperature, water exposure, and poor air quality. This work must by conducted at great speeds and with constant awareness of personal danger. No wonder there is such concern for hiring potential firefighters who are physically and emotionally suited for the job.

Federal law prohibits discrimination against job applicants on the basis of age. However, when youth is a bona fide prerequisite for effective performance of the duties of the job, employers are permitted to prescribe age limits. There is no question that the age limitation on firefighter applicants is justified. The work requires young, agile bodies. There is a big investment of time, effort, and money in the training of firefighters. The goal is to hire and train healthy, capable young men and women who will serve many productive years as effective firefighters.

In fact, medical standards are a matter of official policy:

> This position is physically demanding and affects public health and safety. Therefore, the department requires candidates to meet appropriately high standards of physical and mental fitness. The object of the preemployment examination is to obtain personnel fit to perform in a reasonable manner all the activities of the position. Appointees might be subject to reexamination at any time during their probationary periods.
>
> A candidate will be medically disqualified upon a finding of physical or mental disability that renders the candidate unfit to perform in a reasonable manner the duties of this position or that might reasonably be expected to render the candidate unfit to continue to perform in a reasonable manner the duties of this position. (For example, a latent impairment or a progressively debilitating impairment may, in the judgment of the designated medical officer, reasonably be expected to render the candidate unfit to continue to perform the duties of the position.)
>
> The fitness of each candidate is determined on an individual basis in relation to this position. The designated medical officer may utilize diagnostic procedures including the use of scientific instruments or other laboratory methods that, in his or her discretion, would determine the true condition of the candidate before he or she is accepted. The judgment of the designated medical officer, based on his or her knowledge of the activities involved in the duties of this position and the candidate's condition, is determinative.

The following definitions apply:

- **Physical or Mental Disability:** A physical, mental, or medical impairment resulting from an anatomical, physiological, or neurological condition that prevents the exercise of a normal bodily function or that is demonstrable by medically accepted clinical or laboratory diagnostic techniques. Such impairment may be latent or manifest.

- **Accepted Medical Principles:** Fundamental deduction consistent with medical facts and based on the observation of a large number of cases. To constitute an accepted medical principle, the deduction must be based on the observation of a large number of cases over a significant period of time and be so reasonable and logical as to create a moral certainty that it is correct.
- **Impairment of Function:** Any anatomic or functional loss, lessening, or weakening of the capacity of the body or any of its parts to perform that which is considered by accepted medical principles to be normal.
- **Latent Impairment:** Impairment of function not accompanied by signs and/or symptoms but of such a nature that there is reasonable and moral certainty, according to accepted medical principles, that signs and/or symptoms will appear within a reasonable period of time or upon a change of environment.
- **Manifest Impairment:** Impairment of function accompanied by signs and/or symptoms.
- **Medical Capability:** General ability, fitness, or efficiency (to perform duty) based on accepted medical principles.

As you can see, the medical standards for prospective firefighters are extremely strict and rigid. Conditions that would not deter a person's performance in a less strenuous occupation are absolutely disqualifying for firefighters. Some of these conditions are temporary, such as pregnancy; others, such as the use of drugs or obesity (being severely overweight), might be correctable. Temporary or correctable conditions can cause the candidate to be barred from the physical performance test at a given time and thus might effectively keep the person from competition and employment. Once the temporary or correctable conditions are overcome, however, the candidate can, if still within the age limit, compete at a later date.

A number of medical conditions, depending on their severity, can bar a candidate from employment as a firefighter even if the candidate is able to score high on the physical performance test and is otherwise highly qualified. In fact, a person with any of these conditions might cause him- or herself injury or permanent complications of the condition as a result of undergoing strenuous physical exercise, participating in conditioning programs, or competing in the physical performance test. Some of these conditions include:

- alcoholism
- anemia
- asthma
- chronic gastrointestinal disorders
- diabetes
- drug dependence
- epilepsy or other seizure disorder
- heart disease
- high blood pressure
- joint disease
- kidney disease
- liver disease
- lung disease
- muscular disorders
- obesity (greater than simply being overweight)
- proneness to heat illness such as heat stroke or heat exhaustion
- sickle cell disease
- ulcers

The preceding list represents conditions that are likely to be permanently disqualifying. Other conditions might militate against taking part in physical training or taking the physical performance test until they are removed by passage of time or by active curative measures. Examples of such conditions include:

- acute gastrointestinal disorder
- dehydration
- drug use
- hernia
- infections
- overweight
- pregnancy
- severe underweight

The cooperation and assistance of a physician can help a candidate attain the health and medical status to permit participation in fitness programs and in the test itself. A person who has suffered from any of these conditions or similar ones should remain under the supervision of a doctor while undergoing vigorous physical training and should not attempt to participate in a physical performance test until given approval by the doctor.

Still other physical/medical conditions can disqualify a firefighter candidate simply by virtue of their presence and without regard to physical fitness. Considering the requirements of firefighting, it should be obvious that people with visual impairments (nearsightedness, color blindness, poor depth perception, and so on) cannot function as firefighters. Severe hearing deficit also is disqualifying. Firefighters must not be dependent on glasses or contact lenses or on hearing aids that can be lost, broken, or rendered dysfunctional under heat stress. People with these conditions might be in excellent health and be physically strong and able, but they cannot serve as firefighters. Other medical conditions such as severe allergies, cancer, or the presence of the HIV virus do not augur well for long-term effectiveness as a firefighter, so they are automatically disqualifying to the candidate.

You probably are very much aware of any disabilities or major problems that would keep you from serving as a firefighter. If you are considering yourself as a firefighting candidate, you evidently think of yourself as a healthy, physically fit person. Even so, it would be wise to consult with your doctor before proceeding to invest time and money in the application process and in preparing for the written exam. It also is wise to get the approval of your doctor before you embark on a vigorous physical conditioning program. Tell your doctor about the type of work you have in mind, describe the physical demands, and ask for an assessment of your potential to withstand these rigors. If your doctor foresees any potential problems, either in passing the exams or in facing the demands of the job, discuss corrective measures and remedial programs right now. Follow the medical advice you receive. While speaking with your doctor, you might want to describe the physical performance test. You might be able to pick up special tips to prepare yourself to earn the top score on the performance test itself. Your doctor might even have a physical conditioning program to recommend. Because so much of the hiring decision is based on the physical status of the applicant, it makes sense to devote at least as much attention to preparing your body for the physical test as to preparing your mind for the written exam.

The exam for which you are preparing yourself is the first step toward your career in firefighting. If all goes well, you are planning to spend the bulk of your working life in this career. You cannot expect to prepare in a few weeks for a career that will last a lifetime. Plan ahead. Start now.

The book you have before you is a good starting place for preparation for the written test. Now consider preparing for the physical test at the same time.

Step one: Go to your doctor for a full medical exam. If you get the go-ahead from your doctor, you can begin an aggressive physical fitness training program. If your doctor has any reservations, take care of your health problems. Follow your doctor's advice for improving your general health, for curing any specific problems, and for building strength and stamina at a rate consistent with your overall condition.

Step two: Stop polluting your body. If you smoke, stop smoking now. Medical authorities do not agree on all the details, but they all acknowledge that some harm is caused by tobacco smoke. Smoking has been linked to lung cancer, breathing disorders, heart disease, and circulatory diseases. Smoking begins to affect the body long before actual diseases develop. Smokers generally do not have the stamina of nonsmokers. Smoking reduces the power of the lungs to distribute oxygen. With less oxygen going to the muscles of the body, smokers cannot run or climb as long, cannot carry as much, cannot sustain effort to the degree that nonsmokers can. For a top score on the physical performance test and for a long career in firefighting, stop smoking now.

If you use drugs—inhaled, injected, ingested, or smoked—stop now. Different drugs affect the body in different ways, but all alter functioning in some way. Regardless of whether the drug you use affects your vision, your hearing, your emotions, or your thinking, it creates an unnatural condition that does not enhance your chances of top scores on examinations nor your chances of being selected for a firefighter position. In fact, drug testing is part of the screening process for firefighters. People who test positive for drug use cannot be hired. Medical authorities warn that the intense activity of the physical performance test can cause permanent damage to the body of a drug user.

Excess alcohol and excess caffeine also are body pollutants. Moderate amounts of these stimulants probably are not harmful, though experts disagree as to the definition of "moderate." Let common sense be your guide. Do not overdo it.

Step three: Control your weight. Excess weight requires your muscles, heart, and lungs to work harder. An overweight person tends to have less speed and less stamina than a well-proportioned person and tends to have a shorter productive life span as well. Rapid weight reduction, however, constitutes a shock to the system. Quick weight loss is unhealthy. Weight must be lost gradually through a combination of calorie cutback and sensible exercise.

Very low body weight also can be a problem for the prospective firefighter. A person who is severely underweight is unlikely to be able to carry heavy loads for any length of time and may tire easily. If being underweight does not stem from disease, an underweight person should be able to gain slowly by eating more food and by adding muscle weight through exercise and weight lifting.

Each department sets its own standards, but the following tables can serve as a general guide to weight requirements.

ACCEPTABLE WEIGHT IN POUNDS ACCORDING TO FRAME
MALE

A		B	C	D
Height (in bare feet)		Small	Medium	Large
Feet	Inches	Frame	Frame	Frame
5	2	128-134	131-141	138-150
5	3	130-136	133-143	140-153
5	4	132-148	135-145	142-153
5	5	134-140	137-148	144-160
5	6	136-142	139-151	146-164
5	7	138-145	142-154	149-168
5	8	140-148	145-157	152-172
5	9	142-151	148-160	155-176

A Height (in bare feet)		B Small Frame	C Medium Frame	D Large Frame
Feet	Inches			
5	10	144-154	151-163	158-180
5	11	146-157	154-166	161-184
6	0	149-160	157-170	164-188
6	1	152-164	160-174	168-192
6	2	155-168	164-178	172-197
6	3	158-172	167-182	176-202
6	4	162-176	171-187	181-207

ACCEPTABLE WEIGHT IN POUNDS ACCORDING TO FRAME
FEMALE

A Height (in bare feet)		B Small Frame	C Medium Frame	D Large Frame
Feet	Inches			
4	10	102-111	109-121	118-131
4	11	103-113	111-123	120-134
5	0	104-115	113-126	122-137
5	1	106-118	115-129	125-140
5	2	108-121	116-132	128-143
5	3	111-124	121-135	131-147
5	4	114-127	124-138	134-151
5	5	117-130	127-141	137-155
5	6	120-133	130-144	140-149
5	7	123-136	133-147	143-163
5	8	126-139	136-150	146-167
5	9	129-142	139-153	149-170
5	10	132-145	142-156	152-173
5	11	135-148	145-159	155-176
6	0	138-151	148-162	158-179

NOTE: Although the preceding tables commence at specified heights, no minimum height requirement has been prescribed. The Civil Service examining physician may determine that weight in excess of that shown in the tables (up to a maximum of 20 pounds) is lean body mass and not fat. Decisions as to frame size of a candidate shall be made by the examining physician.

Step four: Embark on a positive program to build up your strength, agility, speed, and stamina. The chapter that follows will give you some suggestions.

Fitness Tests and Physical Ability Tests

Almost all fire departments require successful candidates to pass some type of physical test. This physical test is called either a fitness test or a physical ability (or agility) test. These two types of tests are very different, yet they are both designed to ensure that individuals are able to perform the physically demanding tasks of the firefighter position.

Fitness tests are designed to measures your level of general "fitness" through sit-ups, mile-and-a-half runs, bench press repetitions, and so on. The fitness test uses different cutoff scores in these components for men and women and for various age groups; however, these cutoff scores are set at the same percentile for everyone. In other words, even if men and women are not required to bench press the same amount of weight, the amount required of each can probably be pressed by about three fourths (75 percent) of the men and three fourths (75 percent) of the women.

Physical ability (or agility) tests require candidates to perform a series of linked exercises that simulate a firefighter's job such as swinging an ax, climbing stairs with equipment, and dragging a dummy. This type of physical test is not measuring general fitness. Rather, it is measuring your ability to perform essential job tasks. Because the tasks are essential to the job, the same score (the ability to perform these tasks in a reasonable period of time) is required of all candidates to pass regardless of sex or age. Physical ability (or agility) tests also require a high level of physical fitness to perform well.

DEFINING FITNESS

Physical fitness is to the human body what fine tuning is to an engine. It enables us to perform up to our potential. Fitness can be described as a condition that helps us look, feel, and do our best. More specifically, it is:

> "The ability to perform daily tasks vigorously and alertly, with energy left over for enjoying leisure-time activities and meeting emergency demands. It is the ability to endure, to bear up, to withstand stress, to carry on in circumstances where an unfit person could not continue and is a major basis for good health and well-being."

Physical fitness involves the performance of the heart and lungs and the muscles of the body. Because what we do with our bodies also affects what we can do with our minds, fitness influences to some degree qualities such as mental alertness and emotional stability.

As you undertake your fitness program, it's important to remember that fitness is an individual quality that varies from person to person. It is influenced by age, sex, heredity, personal habits, exercise, and eating practices. You can't do anything about the first three factors, but it is within your power to change and improve the others as necessary.

Knowing the Basics

Physical fitness is most easily understood by examining its components or parts. There is widespread agreement that these four components are basic:

CARDIORESPIRATORY ENDURANCE—The ability to deliver oxygen and nutrients to tissues and to remove wastes over sustained periods of time. Long runs and swims are among the methods employed in measuring this component.

MUSCULAR STRENGTH—The ability of a muscle to exert force for a brief period of time. Upper-body strength, for example, can be measured by various weight-lifting exercises.

MUSCULAR ENDURANCE—The ability of a muscle, or a group of muscles, to sustain repeated contractions or to continue applying force against a fixed object. Pushups often are used to test endurance of arm and shoulder muscles.

FLEXIBILITY—The ability to move joints and use muscles through their full range of motion. The sit-and-reach test is a good measure of flexibility of the lower back and the backs of the upper legs.

BODY COMPOSITION—This refers to the makeup of the body in terms of lean mass (muscle, bone, vital tissue, and organs) and fat mass. An optimum ratio of fat to lean mass is an indication of fitness, and the right type of exercises can help you decrease body fat and increase or maintain muscle mass.

A WORKOUT SCHEDULE

How often, how long, and how hard you exercise and what types of exercises you do should be determined by what you are trying to accomplish. Right now, your goal is to prepare your body so it can withstand the rigors of the physical performance test and make a top showing. Your goal also should be solid preparation for the physical demands of firefighting.

Your exercise program should include something from each of the four basic fitness components previously described. Each workout should begin with a warm up and should end with a cool down. As a general rule, space your workouts throughout the week and avoid consecutive days of hard exercise.

The following is the amount of activity necessary for the average, healthy person to maintain a minimum level of overall fitness. Included are some of the popular exercises for each category.

WARM UP—5–10 minutes of exercises such as walking, slow jogging, knee lifts, arm circles, or trunk rotations. Low-intensity movements that simulate movements to be used in the activity also can be included in the warm up.

MUSCULAR STRENGTH—a minimum of two 20-minute sessions per week that include exercises for all the major muscle groups. Lifting weights is the most effective way to increase strength.

MUSCULAR ENDURANCE—at least three 30-minute sessions each week that include exercises such as calisthenics, pushups, sit-ups, pull-ups, and weight training for all the major muscle groups.

CARDIORESPIRATORY ENDURANCE—at least three 20-minute bouts of continuous aerobic (activity requiring oxygen), rhythmic exercise each week. Popular aerobic conditioning activities include brisk walking, jogging, swimming, cycling, rope-jumping, rowing, cross-country skiing, and some continuous-action games such as racquetball and handball.

FLEXIBILITY—10–12 minutes of daily stretching exercises performed slowly and without a bouncing motion. This can be included after a warm up or during a cool down.

COOL DOWN—a minimum of 5–10 minutes of slow walking or low-level exercise combined with stretching.

A Matter of Principle

The keys to selecting the right types of exercises for developing and maintaining each of the basic components of fitness are found in these principles:

SPECIFICITY—pick the right type of activities to affect each component. Strength training results in specific strength changes. In addition, train for the specific activity in which you're interested. For example, optimum swimming performance is best achieved when the muscles involved in swimming are trained for the movements required. It does not follow that a good runner is a good swimmer.

OVERLOAD—work hard enough, and at levels that are vigorous and long enough to overload your body above its resting level, to bring about improvement.

REGULARITY—you can't hoard physical fitness. At least three balanced workouts per week are necessary to maintain a desirable level of fitness.

PROGRESSION—increase the intensity, frequency, and/or duration of activity over periods of time to improve.

Some activities can be used to fulfill more than one of your basic exercise requirements. In addition to increasing cardiorespiratory endurance, for example, running builds muscular endurance in the legs, and swimming develops the arm, shoulder, and chest muscles. If you select the proper activities, it is possible to fit parts of your muscular endurance workout into your cardiorespiratory workout and save time.

Measuring Your Heart Rate

Heart rate is widely accepted as a good method for measuring intensity during running, swimming, cycling, and other aerobic activities. Exercise that doesn't raise your heart rate to a certain level and keep it there for 20 minutes won't contribute significantly to cardiovascular fitness.

The heart rate you should maintain is called your **target heart rate.** There are several ways to arrive at this figure. One of the simplest is as follows:

> **maximum heart rate** = (220 – age)
> **target heart rate** = maximum heart rate × 70%

Thus, the target heart rate for a 40-year-old individual would be 126.

Some methods for figuring the target rate take individual differences into consideration. Here is one such method:

1. Subtract your age from 220 to find your maximum heart rate.
2. Subtract your resting heart rate (see below) from your maximum heart rate to determine your heart rate reserve.
3. Take 70 percent of your heart rate reserve to determine your heart rate raise.
4. Add your heart rate raise to your resting heart rate to find your **target rate.**

Your resting heart rate should be determined by taking your pulse after sitting quietly for five minutes. When checking heart rate during a workout, take your pulse within five seconds after interrupting exercise because it starts to go down once you stop moving. Count your pulse for 10 seconds and then multiply by six to get the per-minute rate.

A Measure of Your Progress

You will be able to observe the increase in your strength and stamina from week to week in many ways.

There also is a two-minute step test you can use to measure and keep a running record of the improvement in your circulatory efficiency, one of the most important of all aspects of fitness.

The immediate response of the cardiovascular system to exercise differs markedly between well-conditioned individuals and others. The test measures the response in terms of pulse rate taken shortly after a series of steps up and down onto a bench or chair.

Although it does not take long, it is necessarily vigorous. Stop if you become overly fatigued while taking the step test. You should not try it until you have completed a number of weeks of conditioning exercises.

The Step Test

Use any sturdy bench or chair 15–17 inches in height.

> Count 1—Place right foot on bench.
> Count 2—Bring left foot alongside of right and stand erect.
> Count 3—Lower right foot to floor.
> Count 4—Lower left foot to floor.
> REPEAT the four-count movement 30 times a minute for two minutes.
> THEN sit down on a bench or chair for two minutes.
> FOLLOWING the two-minute rest, take your pulse for 30 seconds.
> Double the count to get the per-minute rate. (You can find your pulse by applying the middle and index finger of one hand firmly to the inside of the wrist of the other hand, on the thumb side.)

Record your score for future comparisons. In succeeding tests—about once every two weeks—you probably will find your pulse rate becoming lower as your physical condition improves.

Three important points:

1. For best results, do not engage in physical activity for at least 10 minutes before taking the test. Take it at about the same time of day and always use the same bench or chair.
2. Remember that pulse rates vary among individuals. This is an individual test. What is important is not a comparison of your pulse rate with that of anybody else but rather a record of how your own rate is reduced as your fitness increases.
3. As you progress, the rate at which your pulse is lowered should gradually level off. This is an indication that you are approaching peak fitness.

STAYING FIT

When you have reached the level of conditioning you have chosen for yourself, you will want to maintain your fitness.

Although it has been found possible to maintain fitness with three workouts a week, exercise ideally should be a daily habit. If you can, by all means continue your workouts on a five-times-a-week basis.

Broadening Your Program

There are many other activities and forms of exercise that, if you want, you can use to supplement a basic program.

These activities include a variety of sports, water exercises you can use if you have access to a pool, and isometrics—sometimes called exercises without movement—which take little time (6–8 seconds each).

Isometrics

Isometric contraction exercises take very little time and require no special equipment. They're excellent muscle strengtheners and, as such, are valuable supplements.

The idea of isometrics is to work out a muscle by pushing or pulling against an immovable object such as a wall or by pitting it against the opposition of another muscle.

The basis is the "overload" principle of exercise physiology, which holds that a muscle required to perform work beyond the usual intensity will grow in strength. Research has indicated that one hard, 6- to 8-second isometric contraction per workout can, over a period of six months, produce a significant strength increase in a muscle.

The exercises illustrated and described in the following pages cover the major large muscle groups of the body.

They can be performed almost anywhere and at almost any time.

There is no set order for doing them nor do they all have to be completed at one time. You can, if you like, do one or two in the morning and others at various times during the day whenever you have half a minute or even less to spare.

For each contraction, maintain tension for *no more than 8 seconds*. Do little breathing during a contraction; breathe deeply between contractions.

Start easily. Do *not* apply maximum effort in the beginning.

For the first three or four weeks, you should exert only about one-half of what you think is your maximum force.

Use the first three or four seconds to build up to this degree of force and use the remaining four or five seconds to hold it.

For the next two weeks, gradually increase force to more nearly approach maximum. After about six weeks, it will be safe to exert maximum effort.

Pain indicates that you're applying too much force; reduce the amount immediately. If pain continues to accompany any exercise, discontinue using that exercise for a week or two. Then try it again with about 50 percent of maximum effort; if no pain occurs, you can go on to gradually build up toward maximum.

Neck

Starting position: Sit or stand with interlaced fingers of your hands on your forehead.

Action: Forcibly exert a forward push of your head while resisting equally hard with your hands.

Starting position: Sit or stand with interlaced fingers of your hands behind your head.

Action: Push your head backward while exerting a forward pull with your hands.

Starting position: Sit or stand with the palm of your left hand on the left side of your head.

Action: Push with your left hand while resisting with your head and neck. Reverse using your right hand on the right side of your head.

Upper Body

Starting position: Stand with your back to the wall, your hands at your sides, your palms toward the wall.

Action: Press your hands backward against the wall, keeping your arms straight.

Starting position: Stand facing the wall with your hands at your sides and your palms toward the wall.

Action: Press your hands forward against the wall, keeping your arms straight.

Starting position: Stand in a doorway or with your side against a wall, with your arms at your sides and your palms toward your legs.

Action: Press your hand(s) outward against the wall or door frame, keeping your arms straight.

Arms

Starting position: Stand with your feet slightly apart. Flex your right elbow close to your body, palm up. Place your left hand over your right.

Action: Forcibly attempt to curl your right arm upward, giving equally strong resistance with your left hand. Repeat with your left arm.

Arms and Chest

Starting position: Stand with your feet comfortably spaced and your knees slightly bent. Clasp your hands, palms together and close to your chest.

Action: Press your hands together and hold.

Starting position: Stand with your feet slightly apart and your knees slightly bent. Grip your fingers with your arms close to your chest.

Action: Pull hard and hold.

Abdominal

Starting position: Stand with your knees slightly flexed and your hands resting on your knees.

Action: Contract your abdominal muscles.

Lower Back, Buttocks, and Back of Thighs

Starting position: Lie face down with your arms at your sides, your palms up, and your legs placed under a bed or another heavy object.

Action: With both hips flat on the floor, raise one leg, keeping your knee straight so that the heel pushes hard against the resistance. Repeat with opposite leg.

Legs

Starting position: Sit in chair with your left ankle crossed over your right, with your feet resting on floor and your legs bent at 90-degree angle.

Action: Forcibly attempt to straighten your right leg while resisting with the left. Repeat with opposite leg.

Inner and Outer Thighs

Starting position: Sit with your legs extended, with each ankle pressed against the outside of sturdy chair legs.

Action: Keep your legs straight and pull toward one another firmly. For outer thigh muscles, place your ankles inside chair legs and exert pressure outward.

Water Activities

Swimming is one of the best physical activities for people of all ages—and for many people who are handicapped.

With the body submerged in water, blood circulation automatically increases to some extent. The pressure of water on the body also helps promote deeper ventilation of the lungs. With well-planned activity, both circulation and ventilation increase still more.

Weight Training

Weight training also is an excellent method of developing muscular strength—and muscular endurance. Both barbells and weighted dumbbells—complete with instructions—are available at most sporting goods stores. A good rule to follow in deciding the maximum weight you should lift is to select a weight you can lift six times without strain. If you have access to a gym with sophisticated equipment, take advantage of the advice of professional trainers in establishing weight training programs and goals.

Sports

Soccer, basketball, handball, squash, ice hockey, and other sports that require sustained effort can be valuable aids in building circulatory endurance.

If you have been sedentary, however, it's important to pace yourself carefully in such sports. It might even be advisable to avoid them until you are well along in your physical conditioning program. This doesn't mean you should avoid all sports.

There are many excellent conditioning and circulatory activities in which the amount of exertion can be easily controlled and in which you can progress at your own rate. Bicycling is one example. Others include hiking, skating, tennis, running, cross-country skiing, rowing, canoeing, water skiing, and skin-diving.

Games should be played with full speed and vigor only when your conditioning permits doing so without undue fatigue.

On days when you get a good workout in sports, you can skip part or all of your exercise program. Use your own judgment.

If you have engaged in a sport that exercises the legs and that stimulates the heart and lungs—such as skating—you could skip the circulatory activity for that day, but you still should do some of the conditioning and stretching exercises for the upper body. On the other hand, weight-lifting is an excellent conditioning activity, but it should be supplemented with running or one of the other circulatory exercises.

Whatever your favorite sport, you will find your enjoyment enhanced by improved fitness. Every weekend athlete should invest in frequent workouts.

District of Columbia Fire and Emergency Medical Services Department Firefighter Physical Ability Test

INTRODUCTION

This test consists of a series of tasks designed to assess important physical abilities necessary for effective job performance as an entry-level firefighter. These tasks were developed to mirror real situations that firefighters might encounter on the job.

In developing the Physical Ability Test, a group of firefighter experts from the D.C. FEMSD identified many of the tasks essential to the performance of the job of a firefighter. Job analysis questionnaire data also was collected from incumbent firefighters; this provided the background information necessary to develop this job-related physical ability test.

From this information, nine job-related Physical Ability Test components were developed. These nine components are:

- aerial ladder climb
- hydrant opening
- ladder carry
- ladder extension
- charged line advance
- stair climb with equipment
- ceiling pole
- victim rescue in a confined space
- victim rescue dummy drag

Important Information About the Test

The physical ability test consists of a series of events designed to simulate such firefighter-related activities as carrying fire equipment up flights of stairs, opening fire hydrants, rescuing a victim while in a confined space, and carrying and extending a ladder. The following points should help applicants familiarize themselves with what will take place on the day of the physical ability test.

- Applicants must wear a turnout coat, a hard hat, and a completely filled self-contained breathing apparatus (SCBA and composite bottle, excluding the face piece and low-pressure hose) for all but the first test component (the Aerial Ladder Climb). This equipment will be supplied to the applicants at the test site. Applicants should wear athletic shoes, and it is recommended that they wear long pants for safety.

- Timing of the testing components is to begin *after* the first event (the Aerial Ladder Climb) has been successfully completed. Applicants unable to successfully complete the aerial ladder climb will not be allowed to continue the testing process.
- All remaining testing components must be completed in a series. There are no breaks between timed events.
- During the sequence of timed events, applicants will be permitted to run, walk, or rest between testing stations if they choose to do so. Given the extremely demanding nature of the test, it is recommended that applicants pace themselves so they can complete the entire series of events.
- Prior to the actual examination, a test administrator will give an orientation and thorough walk-through to all candidates. During the walk-through, applicants will not be permitted to touch any of the equipment. Applicants will be given a demonstration of each of the test components.
- One test monitor will time candidates as they proceed through the test stations. Additional monitors will be located at each test station to provide direction to candidates and to observe their health and safety.
- The Physical Ability Test is timed from beginning to end and must be completed within a specific period of time to pass.

DESCRIPTION OF THE PHYSICAL ABILITY TEST

The following nine physical tasks compose the entire Physical Ability Test. Each task reflects the types of situations that an entry-level firefighter might encounter while on the job.

Untimed Event

1. <u>Aerial Ladder Climb:</u> For this first event, applicants must climb to the top of an aerial ladder set at 60 degrees to a height of 50 feet and then return to the bottom. Applicants will be attached to a safety line and will, upon instruction, ascend and descend the ladder without stopping. This is the first event of the Physical Ability Test, and it is not timed. Applicants unable to successfully complete this event will not be allowed to continue the testing process. For this event only, applicants will not be required to wear the turnout coat or the SCBA; however, they will be required to wear the hard hat for safety.

Timed Events

2. <u>Hydrant Opening</u>: Applicants must use a hydrant wrench to open a functional hydrant completely (17 turns) and to close it completely (17 turns).
3. <u>Ladder Operations—Carry</u>: Applicants must remove a 12-foot plain ladder from a truck and carry it a distance of 50 feet, placing it against the drill tower. The ladder must then be picked up, carried back, and replaced onto the truck in its original position.
4. <u>Ladder Operations—Extension</u>: Applicants must extend a 30-foot ground ladder that is affixed to the side of the drill tower and bring it back down. The ladder cannot be dropped.
5. <u>Charged Line Advance</u>: Applicants must pick up and advance a charged $1\frac{1}{2}$-inch hoseline with playpipe attached for a distance of 100 feet. Pump pressure is set at 130 psi.

6. <u>Stair Climb with Equipment</u>: Applicants must pick up a 1½-inch standpipe rack (a wrapped hose) and carry it up six flights of stairs. At the top of the stairs, applicants must place the standpipe rack on the floor. Applicants must then pick up the standpipe rack again and carry it back down to the bottom of the stairs. Note that this is an extremely difficult test component given the weight of the equipment worn and carried.
7. <u>Ceiling Pole</u>: Simulators have been constructed to replicate the action involved when using a ceiling pole to break apart ceilings in structures. Applicants must complete *two sets* of repetitions on *both* ceiling pole simulators (up and down). There are four total sets. Each set consists of six repetitions.
8. <u>Victim Rescue—Confined Space</u>: Applicants must wear a blackened facepiece to block their vision and will be directed to enter a maze. Applicants must navigate the maze on their hands and knees and must exit at the opposite end.
9. <u>Victim Rescue—Dummy Drag</u>: Applicants must drag a 165-pound human-form dummy for a distance of 50 feet.

The amount of time needed to successfully pass the test is based on the times of incumbent firefighters who proceeded through the course at a pace they used while performing these tasks on the job.

THE LAYOUT OF THE TESTING SITE

The diagram at the end of this information guide illustrates the layout of the physical ability testing sequence. Each testing component is labeled with a number that corresponds to the components described in the preceding section.

CLOTHING WORN AND EQUIPMENT USED

The following is a list of clothing to be worn by candidates during the physical ability test and equipment that will be used. Note the weight of the various pieces of equipment:

Equipment	Weight
Turnout coat	5.5
SCBA (filled)	34
12 foot ladder	27
1½-inch standpipe rach	57
Ceiling pole simulator (up)	75
Ceiling pole simulator (down)	75
Human form dummy	165

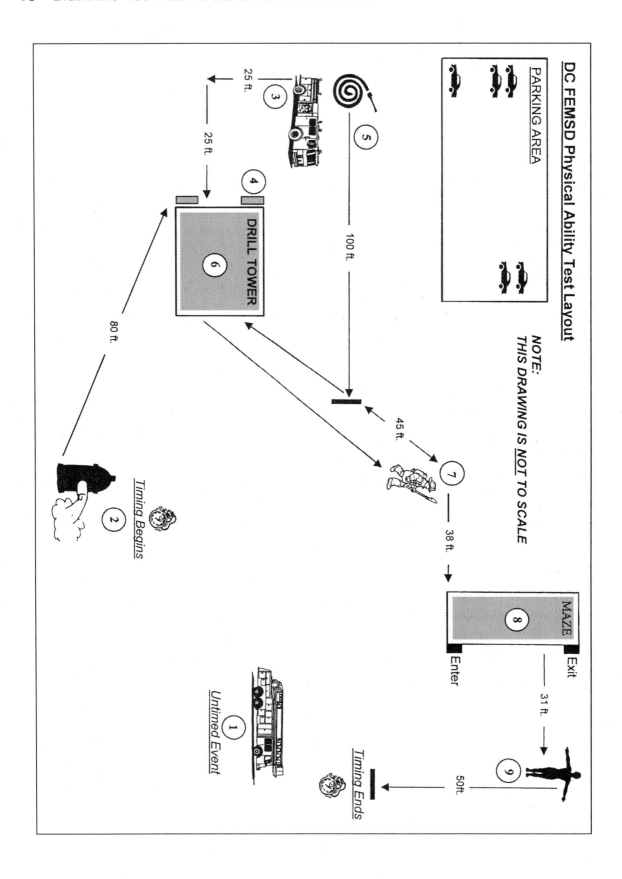

Illinois Fire and Police Recruitment Administration, Inc. Personnel Aptitude Screening Test

ILLINOIS FIRE & POLICE
RECRUITMENT ADMINISTRATION, INC.
PERSONNEL APTITUDE SCREENING SERVICE

FIREFIGHTER PHYSICAL FITNESS TEST

How Will the Physical Ability Be Measured?

The firefighter physical ability test consists of four pass-required events and four additional events that are assessments used for evaluation only. Each event is a scientific and valid test. The test will be given in sequence with a rest period between each event.

The required performance to pass each event is based on sex and age (decade). Although the absolute performance is different for each category, the relative level of effort is identical for each age and sex group. All candidates are required to meet the same percentile rank in terms of their respective age and sex groups. The performance requirement is the level of physical performance that approximates the 40th percentile for each age and sex group.

Test	Male				Female			
	20-29	30-39	40-49	50-59	20-29	30-39	40-49	50-59
Sit and Reach	16.0"	15.0"	13.8"	12.8"	18.8"	17.8"	16.8"	16.3"
Minute Sit-Up	37	34	28	23	31	24	19	13
Bench press	0.98	0.87	0.79	0.7	0.58	0.52	0.49	0.43
1.5 mile run	13.46 min	14.31 min	15.24 min	16.21 min	16.21 min	16.52 min	17.53 min	18.44 min

I. One Minute Sit-Up Test

This is a measure of the muscular endurance of the abdominal muscles. It is an important area for performing fire tasks that might involve the use of force, and it is an important area for maintaining good posture and for minimizing lower-back problems. The score is the number of bent-leg sit-ups performed in one minute.

II. Sit and Reach Test

This is a measure of the flexibility of the lower-back and upper-leg areas. It is an important area for performing fire tasks involving range of motion, and it is important in minimizing lower-back problems. The test involves stretching out to touch the toes or beyond with extended arms from the sitting position. The score is in the inches reached on a yard stick with 15 inches representing the toes.

III. Maximum Bench Press (One Repetition)

This is a maximum weight pushed from the bench press position. It measures the amount of force the upper body can generate. It is an important area for performing fire tasks requiring upper-body strength. The score is a ratio of weight pushed divided by body weight.

IV. One and One Half (1.5) Mile Run

This is a timed run to measure the heart and vascular systems' capability to transport oxygen. It is an important area for performing fire tasks involving stamina and endurance, minimizing the risk of cardiovascular problems. The score is in minutes and seconds. This run is measured using a treadmill within a controlled atmosphere. (We hold the option of testing on treadmill or an indoor or outdoor track.)

V. Threshold Weight and Skinfold Body Fat Percentage

- This is not a disqualifying event -

This is the weight that has been determined as the weight necessary to perform firefighting tasks without undue effort and to minimize health problems due to excessive body fat. The weight should not be more than the recommended pounds per height in inches.

HT/IN	Threshold weight	HT/IN	Threshold weight	HT/IN	Threshold weight
52	75	63	134	74	217
53	80	64	141	75	226
54	85	65	147	76	235
55	89	66	154	77	245
56	94	67	161	78	255
57	99	68	168	79	265
58	105	69	176	80	275
59	110	70	184	81	285
60	116	71	192	82	297
61	121	72	200	83	307
62	128	73	209	84	318

VI. Carry and Balance Test Assessment

- This is not a disqualifying event -

This test is designed to measure how well the joint and muscle sensors react to control movement and maintain balance.

Procedure: The candidate, given an oblong object weighing at least 20 pounds, will walk the length of a beam (measuring more than 12 feet long by 3–4 inches wide) secured to a level floor without falling off or stepping off the beam.

VII. Lumbar Spine Endurance Assessment

- This is not a disqualifying event -

This test is designed to measure how well the lumbar spine will react under strenuous activities and movement.

Procedure: Holding an object weighing at least 5 pounds, standing upright, and keeping feet within a 12-inch area, the candidate will bend down and touch the floor with the weighted object at least 12 inches to the left of the left foot, return to the upright position, and then do the same on the right side. This counts as one repetition. The score is a minimum of 20 repetitions in one minute.

VIII. Five-Minute Claustrophobia Assessment (Covered Mask)
- This is not a disqualifying event -

This event is designed to measure how the candidate will react in a closed, tightly sealed area. It is important for performing fire tasks involving face masks in hazardous environments.

Procedure: Given a covered mask, the candidate will secure a mask to his or her face for a minimum of five minutes without interruption.

PERSONALITY TESTING AND PSYCHOLOGICAL EVALUATIONS

Many fire departments include personality testing and/or psychological evaluations as part of their screening process. Personality testing can help departments identify candidates that are really motivated to be firefighters and that have a good attitude about their jobs. Psychological evaluations are much more in-depth and often include both personality testing and an interview with a psychologist. These psychological evaluations are designed to determine a candidate's ability to deal with the extreme stress and other emotionally challenging aspects of being a firefighter. The following are some of the personality characteristics and psychological traits required of a firefighter:

- Stay calm and handle stress
- Maintain a positive attitude
- Maintain emotional control
- Be flexible
- Be decisive
- Be resourceful
- Determine priorities
- Deal with critically injured or ill people
- Handle critical decision-making under life-threatening conditions
- Perform complex tasks under life-threatening conditions
- Work with little or no supervision
- Accept constructive criticism from others
- Take charge when needed
- Counsel, support, and be empathetic toward others
- Perform tasks requiring long periods of intense concentration
- Work under tight timeframes

Because there are no right or wrong answers to questions on personality tests, it is very difficult to prepare for them. There are a few guidelines, however, that can certainly improve your chances of doing well. Keep the following in mind when answering questions on a personality test or when speaking with a psychologist.

Always tell the truth

Personality tests are designed to identify people that are not providing candid information. One of the main reasons people do poorly on personality tests is that they answer questions in a way that they think makes them look perfect. This is a big mistake because very few, if any, of us are perfect.

Go with your first thought.

Because there are no right or wrong answers for personality test questions, it's easy to read too much into them. The questions usually are pretty straightforward and should be answered as such. Don't try to second guess the questions or look for hidden meanings; doing so will certainly lower your performance.

Don't be afraid to say how you feel

Many questions on personality tests are answered in terms of how much you agree or disagree with a particular statement. Your choices to such a question might be something like: 1) *Strongly Agree*; 2) *Agree*; 3) *Not Sure*; 4) *Disagree*; 5) *Strongly Disagree*. If you feel strongly about a particular statement, don't be afraid to answer with a 1 or a 5. It's also not a good idea to answer with too many 3's (*Not Sure*) unless you're really not sure about how you feel.

The Interview

An interview can take place at any time during the screening process, but when there are many applicants, the employment interview tends to be the last step. An interview takes a great deal of the interviewer's time, so it usually is offered only to candidates who appear to be fully qualified on the basis of all other measures—written test, background investigation, medical examination, and physical fitness performance test.

If you have passed all the steps up to the interview, you are very close, but you do not yet have a place on the fire force. The interview is a very important step in the firefighter screening process.

The purpose of the interview is to gather information along a number of dimensions and to fill you out as a whole person. Some of the aims of the interview are to:

1. **Supplement the application form.** The interviewer will ask about your childhood, education, and prior employment. He, she, or they will give you an opportunity to explain employment gaps and abrupt terminations. If your record is anything less than perfect, this is your chance to indicate that you have learned from your prior experience. Now is the time to let the interviewer know how you have matured. Do not blame others for your mistakes. Take responsibility for impetuous behavior, personality clashes, and brushes with the law. Make clear that you have developed the self-control to keep from such misbehaviors in the future.

   You also might have an opportunity to expand on the kinds of work you did at prior jobs. If the application form gave you limited space in which to describe duties and responsibilities, you can now fill in details. You can convey enthusiasm for tasks you especially liked or at which you were notably skilled.

2. **Learn of your motivation.** Firefighting is not easy work. The interviewer wants to know why you want to be a firefighter. Is your motivation strictly financial? Do you have an unhappy home life that you want to escape for long stretches? Do you care about people or are you just seeking excitement? How do your interests coincide with those of current firefighters? Do your patterns of interest and motivation match those of successful, satisfied firefighters?

3. **Assess your stability and personality.** This is a tall order, and your interview might or might not do it successfully. The firefighter lives and works under stress. In the firehouse, the firefighter lives in close quarters with other men and women, sharing household and maintenance chores and sacrificing privacy. The firefighter must be able to cooperate with all members of the shift without showing undue irritation at a person who might prove annoying. Likewise, the firefighter must not display habits or mannerisms that might prove irritating to others or create discord in the firehouse. In short, the firefighter must "get along" and must "fit in." The stress that firefighters face is even greater when they fight fires. Interview questions will try to gauge how well you can follow orders and how you will react under conditions of real physical pressure. The interviewer will look for signs as to how you can juggle order-taking with initiative. Some of the questions might be hard to answer. The interview itself is a stressful situation. Consider alternatives and give a decisive answer. A firefighter must think before acting but must not spend too much time thinking about next steps. Try to convey this same balance between deliberation and quick thinking as you answer tough questions. Give the impression that you have the self-confidence both to sustain difficult activity under pressure and to size up situations and deviate from prescribed routine when warranted by extreme emergency. Some fire departments supplement the interview with psychological tests to further screen out candidates who give signs of a tendency to "crack" when in real danger or when faced with frustrating decisions.

Because this interview usually is the final step—the moment of decision as to whether you will be accepted on the force—you will want to come across at your very best. There are bound to be some surprises, but to a large extent, you can prepare yourself for an interview.

First, write down the date, time, and place on a wall calendar at home and in your pocket calendar. Fold the notice of interview into your wallet so you cannot make a mistake. If you are not sure how to get to the interview site, look into public transportation or automobile routes ahead of time. Do a dry run if necessary. On interview day, leave enough time for the unexpected. A three-hour pile-up on the highway will be reported in the newspaper and is an acceptable reason for missing an appointment, but a 20-minute bus breakdown is no excuse. If you arrive much too early, go for a cup of coffee. Err on the side of time to spare. You should go to the floor of your interview about 10 minutes ahead of time so you can go to the restroom to wash your hands and comb your hair.

Next, make sure you dress for the occasion. You are not being interviewed for an executive position. You need not wear a three-piece suit. Your clothing should be businesslike, however, clean and neat. You should be well groomed. You want to impress the interviewer as someone he or she would find pleasant to spend a lot of time with. You will add to this impression with a firm handshake, an easy smile, and frequent eye contact with the interviewer or interviewers. Do not chew gum and do not smoke even if you are offered a cigarette.

Prepare yourself with answers to various questions. You might wonder how you can prepare the answers to questions you have not seen. Actually, this is easy. The questions are quite predictable. Begin your preparation by looking over the application forms you filled out and any other papers you were required to file. You should be able to pick out points that an interviewer will want you to amplify or explain.

Questions that might arise from the information you already have given include:

- Why did you choose your area of concentration in school?
- What particularly interests you about _____ (particular subject)?
- Why did you transfer from x school to y school?
- How did you get the job with _____?
- Which of the duties described in your second job did you like best? Least? Why?
- What did you do during the nine months between your second and third jobs?
- Please explain your attendance pattern at your first job.

- Explain the circumstances surrounding your departure from a particular job.
- Please clarify your armed forces service, arrest record, hospitalization record, and so on, as applicable.

Other questions are almost routine to all interviews. They can be anticipated and prepared for as well.

- Why do you want to leave the kind of work you are doing now?
- Why do you want to be a firefighter?
- How does your family feel about you becoming a firefighter?
- What do you do in your leisure time?
- Do you have hobbies? What are they? What do you particularly like about _____?
- What is your favorite sport? Would you rather play or watch?
- How do you react to criticism? If you think the criticism is reasonable? If you consider the criticism unwarranted?
- What is your pet peeve?
- Name your greatest strengths and weaknesses.
- What could make you lose your temper?
- Of what accomplishment in your life are you most proud?
- What act do you most regret?
- If you could start over, what would you do differently?
- What traits do you value most in a co-worker? In a friend?
- What makes you think you would make a good firefighter?

Still other questions might be more specific to a firefighter interview. You should have prepared answers to the following questions:

- How much sleep do you need?
- Are you afraid of heights?
- What is your attitude toward irregular hours?
- Do you prefer working alone or on a team?
- Are you afraid of dying?
- What would you do with the rest of your life if your legs were crippled in an accident?
- How do you deal with panic? Your own? That of others?
- What is your attitude toward smoking? Drinking? Drugs? Playboy magazine? Gambling?
- What is your favorite TV program? How do you feel about watching the news? Sports? Classical drama? Rock music? Opera? Game shows?

Now make a list of your own. The variety of interview questions is endless, but most can be answered with ease. Preparation makes the whole process much more pleasant and less frightening.

There is one question that strikes terror into the heart of nearly every job candidate. This question is likely to be the first and, unless you are prepared for it, might well throw you off-guard. The question is, "Tell me about yourself." For this question, you should have a prepared script (in your head, not in your hand). Think well ahead of time about what you want to tell. What might the interviewer be interested in? This question is not seeking information about your birth weight or your food preferences. The interviewer wants you to tell about yourself with relation to your interest in and qualifications for firefighting. Think about how to describe yourself with this goal in mind. What information puts you in a good light with reference to the work for which you are applying? Organize your presentation. Analyze what you plan to say. What is an interviewer likely to pick up on? To what questions will your little speech lead? You must prepare to answer these questions to which you have opened yourself.

The temptation to talk too much is greatest with this open-ended question, but it exists with many other questions as well. Resist the urge to ramble on and on. Give full, complete answers but stop when you have answered the questions. Do not tell the interviewer more than he or she needs to know. Do not volunteer extra, superfluous information. Do not get anecdotal, chatty, or familiar. Remember that the interview is a business situation, not a social one.

Toward the end of the interview, the interviewer most likely will ask if you have any questions. You undoubtedly will have had some before the interview and should come prepared to ask. If all your questions have been answered during the course of the interview, you might tell this to the interviewer. If not, or if the interview has raised new questions in your mind, by all means ask them. The interview should serve for your benefit as well; it is not just to serve the purposes of the Personnel Division or the Fire Department.

The invitation of your questions tends to be the signal that the interview is nearly over. The interviewer is satisfied that he or she has gathered enough information. The time allotted to you is up. Be alert to the cues. Do not spoil the good impression you have made by trying to prolong the interview.

At the end of the interview, smile, shake hands, thank the person for the opportunity to have the interview, and leave. A brief thank-you note to the interviewer or to the chairperson of the interviewing panel would not be out of order. Such a note brands you as a courteous, thoughtful person and indicates your continued interest in appointment.

GLOSSARY OF FIREFIGHTING TERMINOLOGY

access stairs—In a high-rise building, a stairway, usually open, that serves a number of floors of a common tenant. Also known as convenience stairs.

aerial ladder—A large, metal extension ladder mounted on a truck. It has a bed ladder with three sliding sections, and the overall length is generally 100 feet.

apparatus—The term given to the rolling stock of a fire department such as a ladder truck, pumpers, and rescue vans.

attack stairway—A fire stair used by the fire department to gain access to the fire area. The door between the stairway and the fire area is maintained in an open position.

attic—The space between the ceiling framing of the top story and the underside of the roof.

backdraft—A smoke explosion. If a fire starts in a sealed room, for example, the fire will smolder for lack of oxygen. Heat and fuel are available, but oxygen is lacking. When a door or window is opened, allowing air to rush in, the fire will accelerate with explosive rapidity.

basement—A story partly underground with less than one-half its height below the curb line.

blind shaft elevators—Elevators serving the upper areas of a high-rise building in a shaft that is not equipped with hoistway doors on the lower floors.

bulkhead—An enclosed structure on or above the roof of a building that is not designed or used for human occupancy. For example, a stair bulkhead or an elevator bulkhead.

cellar—A story partly underground with more than one-half its height below the curb line.

charged—The term used to indicate that a hoseline has water under pressure. This term also is used to indicate that an area or building is filled with smoke.

chocks—Wooden wedges carried by firefighters to keep doors open when stretching hoselines.

cockloft—The area in a building between the ceiling of the top floor and the roof boards.

compartmentation—The subdividing of floor areas by fire-resistive separations into smaller spaces or compartments.

core type building—A building in which the elevators, stairways, and building support systems are grouped together in one area of the building. This area could be in the center of the building as in a center core building or on one of the sides of the building as in a side core building.

damper—A device to seal off or control air flow in a heating, ventilation, or air conditioning system.

distributor—A revolving nozzle that is connected to a hose and used for underpier or cellar fires.

drafting—Taking water from a river, a pond, a lake, and so on by suction through a hose.

dropladder—A vertical iron ladder generally found on the first balcony of a fire escape. It should be lowered to the ground before using.

elevating platform, tower ladder, or **cherry picker**—A fire department apparatus used in rescue and firefighting situations. Generally, 75 to 100 feet is the full extension of this equipment.

engine company—The branch of the fire service responsible for the stretching of hoselines and extinguishment of the fire.

evacuation stairway—A fire tower or fire stairs that are remote from the fire area and are used for the evacuation of the building's occupants. A fire tower is the preferred evacuation stairway.

exposures—The structures or areas that border the four sides of the fire.

extinguisher or **"can"**—The "can" is the pressurized water extinguisher used to extinguish fire in its early stages. It generally is carried to the fire by the ladder company first to arrive.

fire damper—A damper used to restrict the passage of heat.

fire safety director—In a high-rise building, a designated employee who holds a certificate of fitness from the fire department qualifying him or her to perform fire safety duties as required.

fire tower—An enclosed stairway connected at each story by an outside balcony or a fireproof vestibule vented to the outside.

forcible entry—Gaining entry to a locked building, store, or apartment using special "forcible entry" tools.

gooseneck—The vertical iron ladder connecting the uppermost fire escape with the roof.

halligan tool—A steel bar approximately $1\frac{1}{4}$ inches thick and 30 inches long with a slightly bent fork at one end and an adz (chisel) and point at the other. It generally is used for forcible entry but has many uses.

hook or **pike pole**—A wooden or fiberglass pole with a steel point and hook. It is used to pull ceilings, to strip plaster from walls, and for many other uses. The poles have lengths of 6, 10, 15, and 20 feet.

hose strap—A short length of strong rope approximately 3 feet long with an iron hook at one end and a loop at the other. It is used to secure a hoseline to a ladder or a fire escape but has many other uses.

ladder company or **truck company**—The branch of the fire service responsible for the raising of ladders, rescue, forcible entry, and ventilation of the fire building.

ladderpipe—A large-caliber nozzle with sufficient hose that is mounted on the upper rungs of an aerial ladder to direct a stream of water into the upper floors of a building.

leeward—The side of a fire that is away from the wind. The smoky side of a fire generally is the leeward side.

locked door fail safe system—A system in which the lock mechanism is controlled electrically from a remote location.

outriggers or **tormentors**—Steel jacks used to stabilize aerial ladders and tower ladders when they are in the elevated position.

overhauling—After a fire has been extinguished, firefighters generally will open walls, floors, and ceilings in the immediate vicinity of the main body of fire to locate any hidden fire.

parapet—The vertical extension of a wall above the roof of a building.

plenum—An air compartment or chamber to which one or more ducts are connected and that forms a part of an air distribution system. In a high-rise building, the space between the suspended ceiling and the underside of the floor above is used as a plenum for the collection of the return air.

"Q" decking—A type of composite floor construction in which corrugated steel is used to support the concrete floor.

reducer—A fitting, generally made of brass or aluminum, that is used to couple hose of two different sizes. A $2\frac{1}{2} \times 1\frac{1}{2}$-inch reducer is used to attach a $1\frac{1}{2}$-inch line to a $2\frac{1}{2}$-inch line.

rolled-up or **donut**—A length of hose that is rolled and tied. Rolled-ups are carried into buildings for firefighters on the upper floors of high buildings.

scissor stairs—Two stairways constructed side by side in the core of a building with doors alternating the point of exit to opposite sides of the core.

scuttle—An opening or hatch in the roof generally covered by a wooden hatch cover.

scuttle ladder—A vertical iron ladder providing access to the roof from the top floor of the building.

search—The action taken by firefighters to find people that might be victims of fire.

sky lobby—An elevator terminal point on an upper floor of a high-rise building where passengers can change from one bank of elevators to another.

skylight—Glass in a metal frame on the roof of a building that allows daylight to enter the room or area. A skylight has louvers that permit ventilation of the room.

stretch—The term used for stretching hoselines at a fire. Example: The lieutenant ordered the firefighters to *stretch* into the fire building.

ventilation—As used in the fire service, ventilation means the opening of the roof, doors, windows, and any other opening in a structure to allow the escape of toxic gases and smoke from a fire.

water hammer—The sudden pressure rise that might accompany a sudden change in the velocity of water flowing in a line. When a nozzle discharging water under pressure is shut down quickly, a water hammer develops that can cause the hose to burst.

windward—The side of the fire that faces into the wind. If you are facing the fire with the wind at your back, you are on the windward side.

READING-BASED QUESTIONS

A recent survey of firefighter examinations given nationwide indicates that there is a wide variation in the subject matter of these exams. The single topic that is common to all exams is reading. Some exams include classic reading-comprehension questions that present a passage and then ask questions about the details of the passage and, perhaps, its meaning. Other exams require candidates to indicate proper behavior based on their reading of printed procedures and regulations. Another type of reading-based question requires candidates to reason and predict next steps on the basis of information presented in a reading passage. Of course, questions of judgment in emergency and non-emergency situations rely heavily on reading as well. Actually, there are nearly as many variations of the reading-based question as there are test-makers.

Before you begin to devote attention to strategies for dealing with reading-based questions, give some thought to your reading habits and skills. Of course, you already know how to read. But how well do you read? Do you concentrate? Do you get the point during your first reading? Do you notice details?

Between now and the test day, you must work to improve your reading concentration and comprehension. Your daily newspaper provides excellent material to improve your reading. Make a point of reading all the way through any article you begin. Do not be satisfied with the first paragraph or two. Read with a pencil in hand. Underscore details and ideas that seem to be crucial to the meaning of the article. Notice points of view, arguments, and supporting information. When you have finished the article, summarize it for yourself. Do you know the purpose of the article? The main idea presented? The attitude of the writer? The points over which there is controversy? Did you find certain information lacking? As you answer these questions, skim back over what you underlined. Did you focus on important words and ideas? Did you read with comprehension?

As you repeat this process day after day, you will find that your reading becomes more efficient. You will read with greater understanding and will "get more" from your newspaper.

You can't sit down the night before a test involving reading comprehension and cram for it. The only way you can build up your reading skill is to practice systematically. The gains you make will show up not only in an increased score on the test but also in your reading for study and pleasure.

Trying to change reading habits you have had for a long time can be difficult and discouraging. Do not attempt to apply *all* the suggestions we have given to *all* your reading *all* at once. Try to follow a program like the following:

1. Set aside 15 minutes each day to practice new reading techniques.
2. Start with a short, easy-to-read article from a magazine or a newspaper. Time yourself. At the end of your practice session, time yourself on another short article and keep a record of both times.
3. Select a news story. Read it first and then practice an eye-scan exercise. Work toward reducing your eye fixations to no more than two per line the width of a newspaper column.
4. Read an editorial, a book review, or a movie or drama review in a literary magazine or newspaper. This type of article always expresses the author's (or the paper's) point of view and therefore is good practice for searching out the main idea. After you read, see whether you can write a good title for the article and jot down in one sentence the author's main idea. You also can try to make up a question based on the article with four choices. This is excellent practice for determining main ideas, and you can use the questions to test your friends.
5. Find one new word and write the sentence in which it appears. Guess at its meaning from the context and then look up the definition in a dictionary. Try to make up a sentence of your own using the word. Then try to use the word in your conversation at least twice the following day.

If you follow this program daily, your test score will show the great gains you have made in reading comprehension.

A major aspect of your daily reading that deserves special attention is vocabulary building. The most effective reader has a rich, extensive vocabulary. As you read, make a list of unfamiliar words. Include in your list words that you understand within the context of the article but that you cannot really define. In addition, mark words you do not understand at all. When you put aside your newspaper, go to the dictionary and look up *every* new and unfamiliar word. Write the word and its definition in a special notebook. Writing the words and their definitions helps seal them in your memory far better than just reading them, and the notebook serves as a handy reference for your own use. A sensitivity to the meaning of words and an understanding of more words will make reading easier and more enjoyable even if none of the words you learn in this way crops up on your exam. In fact, vocabulary building is a good lifetime habit to develop.

Success with reading-based questions depends on more than just reading comprehension. You also must know how to draw the answers from the reading selection and be able to distinguish the best answer from a number of answers that all seem to be good ones or that all seem to be wrong.

Strange as it may seem, it's a good idea to approach reading comprehension questions by reading the questions—not the answer choices, just the questions themselves—before you read the selection. The questions will alert you to look for certain details, ideas, and points of view. Use your pencil. Underscore key words in the questions. These will help you direct your attention as you read.

Next, skim the selection very rapidly to get an idea of its subject matter and its organization. If key words or ideas pop out at you, underline them but do not consciously search out details in the preliminary skimming.

Now read the selection carefully with comprehension as your main goal. Underscore the important words as you have been doing in your newspaper reading.

Finally, return to the questions. Read each question carefully. Be sure you know what it asks. Misreading of questions is a major cause of error on reading comprehension tests. Read *all* the answer choices. Eliminate the obviously incorrect answers. You might be left with only one possible answer. If you find yourself with more than one possible answer, reread the question. Focus on catch words that might destroy the validity of a seemingly acceptable answer. These include expressions such as *under all circumstances, at all times, never, always, under no condition, absolutely, entirely,* and *except when.* Skim the passage once more, focusing on the underlined segments. By now, you should be able to conclude which answer is *best*.

Reading-based questions can take a number of different forms. Some of the most common forms include:

1. **Question of fact or detail.** You might have to mentally rephrase or rearrange, but you should find the answer stated in the body of the selection.
2. **Best title or main idea.** The answer might be obvious, but the incorrect choices to the "main idea" question often are half-truths that are easily confused with the main idea. They might misstate the idea, omit part of the idea, or even offer a supporting idea quoted directly from the text. The correct answer is the one that covers the largest part of the selection.
3. **Interpretation.** This type of question asks you what the selection means, not just what it says. On firefighter exams, the questions based on categories of building styles, for example, might fall into the realm of interpretation.
4. **Inference.** This is the most difficult type of reading-based question. It asks you to go beyond what the selection says and to predict what might happen next. You might have to choose the best course of action to take based on given procedures and a factual situation, or you might have to judge the actions of others. Your answer must be based on the information in the selection and your own common sense

but not on any other information you might have about the subject. A variation of the inference question might be stated as, "The author would expect that…" To answer this question, you must understand the author's point of view and then make an inference from that viewpoint based on the information in the selection.

5. **Vocabulary.** Some firefighter reading sections, directly or indirectly, ask the meanings of certain words as used in the selection.

Let's now work together on some typical reading comprehension selections and questions.

Selection for Questions 1 to 4

The best kind of limestone for printing is Bavarian. Light-colored and perfectly smooth, it is porous and absorbs both water and greasy substances equally well. The stone used is about 6 inches thick and is fairly big, up to 90×65 cm (35×25 inches), and can weigh up to 150 or 175 pounds. Grinding the stone smooth creates a clean printing surface. This allows a drawing to be made on the surface with a greasy lithographic pencil or crayon. The drawing is then fixed by rinsing the stone with a very weak solution of nitric acid and gum arabic. The stone is wiped with water before each impression is taken and, for each print, it is inked with a leather-covered roller. During this process, the porous limestone retains the grease of the crayon where the drawing has been made, and the parts that are not drawn upon become impregnated with water. The ink, which is greasy, is repelled by the water-wet areas and adheres only to the areas marked by the crayon.

1. The best title for the preceding paragraph is
   (A) "Where Good Limestone is Found"
   (B) "The Process of Lithographic Printing"
   (C) "Water Color Drawings on Limestone"
   (D) "How To Make a Printing Stone"

2. According to the preceding paragraph, Bavarian limestone is used because
   (A) it can be cut into very large blocks
   (B) it has a sponge-like surface
   (C) it contains nitric acid and gum arabic
   (D) it will repel grease

3. The aspect of this process NOT described by the preceding paragraph is
   (A) the finished product
   (B) how to ink the stone
   (C) creating a printing surface
   (D) what kind of limestone to use

4. According to the preceding paragraph, the purpose of the nitric acid and gum arabic is
   (A) to make the limestone light-colored
   (B) to erase previous drawings
   (C) to make the drawing permanent
   (D) to repel the printing ink

Begin by skimming the questions and underscoring key words. Your underscored questions should look more or less like this:

1. What is the best title for the preceding paragraph?
2. According to the preceding paragraph, Bavarian limestone is used because
3. The aspect of this process NOT described by the preceding paragraph is
4. According to the preceding paragraph, the purpose of the nitric acid and gum arabic is

Now skim the selection. This quick reading should give you an idea of the structure of the selection and its overall meaning.

Next, read the selection carefully and underscore words that seem important or that you think hold clues to the question's answers. Your underscored selection should look something like this:

> The best kind of limestone for printing is Bavarian. Light-colored and perfectly smooth, it is porous and absorbs both water and greasy substances equally well. The stone used is about 6 inches thick and is fairly big, up to 90×65 cm (35×25 inches), and can weigh up to 150 or 175 pounds. Grinding the stone smooth creates a clean printing surface. This allows a drawing to be made on the surface with a greasy lithographic pencil or crayon. The drawing is then fixed by rinsing the stone with a very weak solution of nitric acid and gum arabic. The stone is wiped with water before each impression is taken and, for each print, it is inked with a leather-covered roller. During this process, the porous limestone retains the grease of the crayon where the drawing has been made, and the parts that are not drawn upon become impregnated with water. The ink, which is greasy, is repelled by the water-wet areas and adheres only to the areas marked by the crayon.

Finally, read the questions and answer choices. Try to choose the correct answer for each question. The correct answers are: 1. **(B)**, 2. **(B)**, 3. **(A)**, 4. **(C)**. Did you get them all right? Whether you made any errors or not, read these explanations:

1. **(B)** The best title for any selection is the one that takes in all the ideas presented without being too broad or too narrow. Choice (B) provides the most inclusive title for this passage. A look at the other choices shows you why. The first sentence mentions where to find good limestone, but the rest of the paragraph clearly indicates that this is not the main idea. Therefore, choice (A) can be eliminated. Water color drawing was not even mentioned, so choice (C) can be eliminated. Although the paragraph emphasizes the importance of the stone, it does not specifically discuss how to create a printing stone. The passage is clearly devoted to the process of creating a print, not how to create a printing stone. Therefore, choice (D) also can be eliminated.

2. **(B)** The capability of Bavarian limestone to absorb water and greasy substances indicates that its sponge-like surface is important for the printing process. Therefore, choice (B) is the best answer. Although the paragraph indicates that the limestone often is cut into very large blocks, nothing in the passage indicates that this is important to the printing process. Therefore, choice (A) can be eliminated. The passage states that the stone is rinsed with a solution of nitric acid and gum arabic. Nothing indicates that the nitric acid and gum arabic come from the stone itself, so choice (C) can be eliminated. The beginning of the passage clearly states that Bavarian limestone absorbs greasy substances. Because the meaning of "repel" is the opposite of "absorb," choice (D) can be eliminated.

3. **(A)** The ideas you underlined should help you with this question of fact. The finished product is never mentioned in the passage. How to ink the stone, how to create a drawing surface, and what kind of limestone to use are all mentioned, so the other three options can be eliminated.

4. **(C)** The answer to this question is provided by the sentence, "The drawing is then fixed by rinsing the stone with a very weak solution of nitric acid and gum arabic." In this sentence, the word "fixed" means "to make permanent." The best way to figure out the meaning of a word used in an unfamiliar way is to decide what meaning would make the most sense based on the rest of the passage. The sentence makes the most sense when we replace the word "fixed" with choice (C) or "made permanent."

Selection for Questions 5 to 9

Weatherproofing wooden deck surfaces with the XYZ Process results in a surface that is far superior to those treated with other, less effective methods. XYZ Process–treated surfaces are durable to heavy walking traffic, are resistant to wear from heavy rains and runoff from overhanging roofs, are resistant to bleaching from excessive exposure to the sun, and maintain a "fresh wood" look for several years. In addition to producing a wooden surface that is attractive and durable, the XYZ Process also has several safety and environmental benefits. Surfaces treated with the XYZ Process are noncombustible, odorless, safe for children and pets, and free from splinters. Any wooden surface can be treated with the XYZ Process and can be treated by anyone with access to a garden hose, a scrub-brush, and a paintbrush. The application process is safe for surrounding vegetation, animal life, and anyone without major respiratory problems.

5. It is most accurate to state that the author in the preceding selection presents
   (A) facts but reaches no conclusion concerning the value of the process
   (B) a conclusion concerning the value of the process unsupported by facts
   (C) neither facts nor conclusions but merely describes the process
   (D) a conclusion concerning the value of the process and facts to support that conclusion

6. If the XYZ Process were used for a surface other than a deck, it would be most useful for
   (A) a cement sidewalk
   (B) a plastic mailbox
   (C) a wooden door
   (D) a metal handrail

7. The aspect of the XYZ Process not discussed in the preceding paragraph is
   (A) resistance to wear from snow and ice
   (B) safety concerns for animals
   (C) effects on surrounding plants
   (D) how it compares to other methods

8. The main reason for treating wooden surfaces with the XYZ Process is to
   (A) protect children from harmful odors
   (B) protect surrounding vegetation
   (C) protect wooden surfaces from weathering
   (D) protect animals from splinters

9. Of the following, the one that might be negatively affected by the XYZ Process is
   (A) plants surrounding the treated area
   (B) someone with respiratory problems
   (C) children walking on a treated surface
   (D) animals walking on a treated surface

9. Ⓐ Ⓑ Ⓒ Ⓓ Ⓔ

Skim the questions and underscore the words you consider to be important. The questions should look something like this:

5. It is most accurate to state that the author in the preceding selection <u>presents</u>
6. If the <u>XYZ Process</u> were used for a <u>surface other than a deck</u>, it would be most useful for
7. The <u>aspect</u> of the XYZ Process not <u>discussed</u> in the preceding paragraph is
8. The <u>main reason for treating</u> wooden surfaces with the XYZ Process is to
9. Of the following, the one that might be <u>negatively affected</u> by the XYZ Process is

Skim the reading selection. Get an idea of the subject matter of the selection and how it is organized.
Now read the selection carefully and underscore the words that you think are especially important. This fact-filled selection might be underlined like this:

<u>Weatherproofing</u> wooden deck surfaces <u>with the XYZ</u> Process results in a surface that is <u>far superior</u> to those treated with other, <u>less effective methods</u>. XYZ Process–treated surfaces are <u>durable to heavy walking traffic</u>, are <u>resistant to wear</u> from heavy rains and runoff from overhanging roofs, are <u>resistant to bleaching</u> from excessive exposure to the sun, and <u>maintain a "fresh wood" look</u> for several years. In addition to producing a wooden surface that is <u>attractive and durable</u>, the XYZ Process also has several <u>safety and environmental benefits</u>. Surfaces treated with the XYZ Process are <u>noncombustible</u>, <u>odorless</u>, <u>safe</u> for <u>children and pets</u>, and free from <u>splinters</u>. <u>Any wooden surface</u> can be treated with the XYZ Process and can be treated by anyone with access to a garden hose, a scrub-brush, and a paintbrush. The application process is <u>safe</u> for <u>surrounding vegetation</u>, <u>animal life</u>, and anyone without <u>major respiratory problems</u>.

Now read each question and all its answer choices. Try to choose the correct answer for each question. The correct answers are: 5. **(D)**, 6. **(C)**, 7. **(A)**, 8. **(C)**, 9. **(B)**.
How did you do? Read the explanations.

5. **(D)** This is a combination main idea and interpretation question. If you cannot easily answer this question, reread the selection. The author clearly states his opinion about the XYZ Process in the first sentence. The rest of the selection provides an abundance of facts to support his conclusion.

6. **(C)** Toward the end of the selection, the author states, "Any wooden surface can be treated with the XYZ Process." Choice (C) is the only one that mentions a wooden surface.

7. **(A)** The ideas you underlined should help you with this factual statement. Resistance to wear from snow and ice is the only aspect not mentioned. Safety concerns for animals, effects on surrounding plants, and how it compares to other methods are all mentioned.

8. **(C)** Protecting wooden surfaces from weathering is the main reason for treating wooden surfaces with the XYZ Process. The fact that the XYZ Process is safe for children, plants, and animals is just another benefit of using this product.

9. **(B)** The only negative effect mentioned is that the XYZ Process might not be safe for people with major respiratory problems.

You should be getting better at reading and answering questions. Try this next selection on your own. Read and underline the questions. Skim the selection. Read and underline the selection. Read the questions and answer choices and mark your answers. Check your answers against the answers and explanations that follow the selection.

Selection for Questions 10 to 12

Most people first encounter grammar in connection with the study of their own language in school. This kind of grammar is called prescriptive grammar because it defines the role of the various parts of speech (such as nouns, verbs, adjectives, and so on). The purpose of prescriptive grammar is to define the norms, or rules, of correct usage. This kind of grammar states how words and sentences are to be put together in a language so the speaker will be perceived as having good grammar. When people are said to have good grammar, the inference is that they obey the rules of accepted usage associated with the language they speak.

10. According to the preceding paragraph, what is the purpose of grammar?
    (A) It is actually not important in our language.
    (B) It is the "glue" that holds sentences together.
    (C) It indicates how to use language properly.
    (D) It obeys the rules of proper language.

11. In the preceding paragraph, the author is primarily concerned with
    (A) the different parts of speech
    (B) the purpose of prescriptive grammar
    (C) who has good grammar
    (D) the study of grammar in school

12. According to the preceding paragraph, someone who is considered to use the rules of prescriptive grammar properly
    (A) has studied their own language in school
    (B) defines the various parts of speech
    (C) did well in language class
    (D) obeys the rules of accepted usage

The correct answers are: 10. **(C)**, 11. **(B)**, 12. **(D)**.

10. **(C)** The paragraph states that the purpose of prescriptive grammar is to define the rules of correct usage. This is just another way of saying that grammar indicates how to use language properly.

11. **(B)** The purpose of this selection should be clear as you read through the paragraph. If you were unsure of the answer, you can easily work backwards by eliminating the wrong answers.

12. **(D)** The answer to this question can be found in the last sentence. The wording of the question is slightly different than the last sentence, but the meaning is the same. If you had difficulty with the meaning of this question, you should have no difficulty answering the question by eliminating the wrong answers.

Now try the next reading selection and its questions. Once more, explanations follow the correct answers. Follow the procedure you have learned and be sure to read the explanations even if you have a perfect score.

Selection for Questions 13 to 18

Information theory deals with the numerical measurement of information, the representation of information (such as encoding), and the capacity of communication systems to transmit, receive, and process information. We use, and may even be a part of, many communication systems on a daily basis. A typical communication system consists of several components. First, there must be a source that produces the information, or message, to be transmitted. A source with which we are familiar is human speech. Second, a transmitter, such as a telephone and an amplifier or a television camera and a transmitter, converts the message into electronic or electromagnetic signals. Third, these signals are transmitted through a channel, or medium, such as a wire or the atmosphere. The channel is the component of the system that is most susceptible to interference. Distortion and degradation of the signal can originate from many sources. (Some examples of interference, known as noise, include the static in radio and telephone reception, the fading of a cellular phone signal, and the "snow" experienced in poor TV reception.) The fourth component of a communication system is the receiver such as a television or a radio. A TV or radio can act as a receiver by reconstructing the signal into the original message. The final component is the destination such as a person watching or listening to the message that the television or radio receives.

13. Which of the following is the best title for the preceding selection?
    (A) "The Basics of Information Theory"
    (B) "Humans and Data Transmission"
    (C) "Communication System Components"
    (D) "Transmission and Signal Interference"

14. According to the preceding selection, the signal is converted by
    (A) the source
    (B) the transmitter
    (C) the receiver
    (D) the channel

15. According to the preceding selection, "degradation of the signal" means the signal
    (A) originates from many sources
    (B) is successfully transmitted
    (C) is graded as a clear signal
    (D) is experiencing interference

16. According to the preceding selection, a human being can act as
   (A) a source only
   (B) a transmitter and a receiver
   (C) a source and a destination
   (D) a destination only

17. According to the preceding selection, what is likely to happen to a telephone signal transmitted through a faulty wire?
   (A) It will result in static.
   (B) It will sound the same.
   (C) It will be amplified.
   (D) It will result in "snow."

18. According to the preceding selection, an example of the representation of information is
   (A) distortion
   (B) encoding
   (C) "snow"
   (D) transmitting

The correct answers are: 13. **(C)**, 14. **(B)**, 15. **(D)**, 16. **(C)**, 17. **(A)**, 18. **(B)**.

13. **(C)** Although this selection starts out with a comment about information theory, this is not the main focus of the selection. Therefore, you can eliminate choice (A). The topics of transmission and interference are discussed but are not the focus of the selection. In the same way, the relationship between humans and this topic is discussed but is not the focus. Therefore, choices (B) and (D) can be eliminated.

14. **(B)** The selection clearly states that the second component of a communication system, the transmitter, converts the message into a signal.

15. **(D)** If you were unsure of the meaning of the word "degradation," you could try to guess its meaning by looking at the context in which the word appears. The sentence containing the word "degradation" appears between two sentences that both discuss interference. In addition, because the word "degradation" is used to describe interference, you can infer that it is just another word for it.

16. **(C)** Human beings are mentioned as possible sources and destinations of a communication system.

17. **(A)** This question requires you to go beyond the available information and guess which answer is most likely to be true given the information provided. A damaged wire is likely to have some negative effects, so choice (B) can be eliminated. Wire damage also is not likely to be beneficial, so choice (C) can be eliminated. The experience of "snow" is specifically described as an effect of interference on TV reception, so choice (D) can be eliminated.

18. **(B)** The first sentence of the selection clearly indicates that encoding is an example of the representation of information.

On firefighter exams, many reading passages will relate to standard designations or to specified rules of procedure. When reading these passages, you must pay special attention to details relating to exceptions, special preconditions, combinations of activities, choices of actions, and prescribed time sequences. Sometimes the printed procedure specifies that certain actions are to be taken only when there is a combination of factors (such as water pressure has dropped AND staircase collapse is imminent). At other times, the procedures give choices of action under certain circumstances. You must read carefully to determine whether the passage requires a combination of factors or gives a choice and then make the appropriate judgment. When a time sequence is specified, be certain to follow that sequence in the prescribed order.

The remaining reading selections in this chapter are of a style often found on firefighter exams.

Beyond requiring routine reading-comprehension skills, they require the special attention to sequence, combinations, and choices we just discussed.

Selection for Questions 19 to 22

Preparing Foods for Storage: Meat, Poultry, and Fish

Always keep your refrigerator at about 40°F and your freezer at 0°F or colder. You should check the temperature periodically with an appliance thermometer. This is very important because, if you do not store these items properly, your customers could become ill from contaminated food. Refrigerate meat and poultry in the original packaging. For long-term freezer storage, it is best to remove the original packaging and wrap it tightly in moisture- and vapor-proof material. You can use heavy foil, freezer bags, freezer plastic wrap, or freezer containers. Tightly wrap fresh fish in moisture- and vapor-proof material before refrigerating or freezing.

19. Assume that you, as the head chef of a major restaurant chain, are responsible for the proper storage of large amounts of meat, poultry, and fish. One of your duties as head chef is to teach your cooks how to store these items. Today, one of your cooks asks you what she should do with a new shipment of fresh fish. You should tell her to
    (A) refrigerate the fish in its original packaging
    (B) store it in the freezer at about 40°F
    (C) freeze the fish in its original packaging
    (D) wrap it in heavy foil and put it in the freezer

20. What the storage instructions mean when they talk about using vapor-proof and moisture-proof materials is
    (A) using materials that do NOT allow liquids or odors to leak out
    (B) using materials that allow liquids but NOT odors to leak out
    (C) using materials that allow odors but NOT liquids to leak out
    (D) using materials that allow odors and liquids to leak out

21. According to the instructions for long-term storage, removing meat or poultry from the original package and wrapping it tightly in freezer plastic wrap is
    (A) inadvisable because you should freeze meat or poultry in the original packaging
    (B) advisable because freezer plastic wrap is moisture- and vapor-proof
    (C) inadvisable because you should not store meat or poultry in the freezer
    (D) advisable because you should not refrigerate meat or poultry in the original packaging

22. According to the storage instructions, the one that would NOT be considered proper food storage is

22. Ⓐ Ⓑ Ⓒ Ⓓ Ⓔ

(A) storing food in a freezer with a temperature that is below 0°F
(B) wrapping meat or poultry in heavy foil before freezing
(C) removing fish from its packaging and storing it in a freezer container
(D) freezing meat or poultry in the original packaging

The correct answers are: 19. **(D)**, 20. **(A)**, 21. **(B)**, 22. **(D)**.

19. **(D)** The last sentence says to wrap fresh fish in moisture- and vapor-proof material. The previous sentence mentions that heavy foil is a vapor-proof material.

20. **(A)** This is somewhat of a vocabulary question. Storage materials that do NOT let odors leak out are called vapor-proof. Storage materials that do NOT let liquids to leak out are called moisture-proof.

21. **(B)** The instructions clearly state that meat and poultry should be removed from their original packaging and wrapped in moisture- and vapor-proof material before long-term storage. Freezer plastic wrap is moisture- and vapor-proof and therefore is suitable for long-term storage.

22. **(D)** The instructions clearly state that meat and poultry should be removed from their original packaging and wrapped in moisture- and vapor-proof material before freezing. Freezing meat or poultry in the original packaging is not considered proper food storage.

Selection for Questions 23 to 26

Citizens interested in filing a property damage claim that totals less than $5,000 are required to fill out Minor Claim Form LG-203 in duplicate and return it to the Secretary of Minor Claims in the District Court Building. If the citizen seeks compensation for emotional damage in addition to the property damage, he or she also is required to fill out Intermediate Claim Form K-46. This form is to be filled out in duplicate and sent to the Secretary of Minor Claims. Minor Claim Form LG-203 will then be compared to Intermediate Claim Form K-46. The Secretary of Minor Claims will then forward both copies of Form K-46 to the Division Office, which will send one copy to the State Office of Minor Claims. When the information on form LG-203 indicates that the claim is of a minor nature, the case is thoroughly reviewed by the Secretary of Minor Claims. If this review indicates that the complaining party is entitled to an emotional-damage hearing, Witness Form R-11 must be completed by a witness to the incident. Once this form is returned to the Secretary of Minor Claims, the defending party is sent a copy of the Summons Form S-4 and is asked to return it to the State Office of Minor Claims. When the State Office of Minor Claims receives the Summons Form, they will compare it to Form K-46 and decide whether the case may be heard in the District Court.

23. According to the preceding paragraph, in order for the State Office of Minor Claims to decide whether a case may be heard in the District Court, they must have received which documents?
- (A) Form K-46
- (B) Forms K-46 and LG-203
- (C) Forms K-46 and S-4
- (D) Forms K-46, LG-203, R-11, and S-4

24. According to the preceding paragraph, the requirement of Witness Form R-11 depends on
- (A) a review by the Secretary of Minor Claims
- (B) whether the citizen does NOT claim emotional damage
- (C) whether the citizen wants to have a witness in court
- (D) a review by the State Office of Minor Claims

25. Of the forms mentioned in the preceding paragraph, the defending party is responsible for preparing the
- (A) Intermediate Claim Form
- (B) Summons Form
- (C) Witness Form
- (D) Minor Claim Form

26. According to the preceding paragraph, the Division Office
- (A) keeps one copy of Form LG-203
- (B) sends out both copies of Form K-46
- (C) keeps one copy of Form K-46
- (D) sends out both copies of Form LG-203

The correct answers are: 23. **(C)**, 24. **(A)**, 25. **(B)**, 26. **(C)**.

23. **(C)** The State Office of Minor Claims must receive Form K-46 from the Division Office and Form S-4 from the defending party.

24. **(A)** If a review by the Secretary of Minor Claims indicates that the complaining party is entitled to an emotional-damage hearing, Witness Form R-11 must be completed.

25. **(B)** The defending party is only responsible for the Summons Form.

26. **(C)** The Division Office receives both copies of Form K-46 and sends one copy to the State Office of Minor Claims. This implies that the Division Office keeps one copy of Form K-46.

READING COMPREHENSION FOR FIREFIGHTERS

Directions: Questions 1 through 4 relate to the following paragraph and are to be answered in accordance with the paragraph.

The unadjusted loss per $1,000 valuation has only a very limited usefulness in evaluating the efficiency of a fire department, for it depends on the assumption that other factors will remain constant from time to time and city to city. It might be expected that high fire department operation expenditures would tend to be associated with a low fire loss. A statistical study of the loss and cost data in more than 100 cities failed to reveal any such correlation. The lack of relationship, although to some extent due to failure to make the most efficacious expenditure of fire protection funds, must be attributed at least in part to the obscuring effect of variations in the natural, physical, and moral factors that affect fire risk.

1. One reason for the failure to obtain the expected relationship between fire department expenditures and fire loss data is
    (A) the changing dollar valuation of property
    (B) the unsettling effects of rapid technological innovations
    (C) the inefficiency of some fire department activities
    (D) the statistical errors made by investigators

2. We can conclude that the "unadjusted loss per $1,000" figure is useful in comparing the fire departments of two cities
    (A) only if the cities are of comparable size
    (B) only if adjustments are made for other factors that affect fire loss
    (C) under no circumstances
    (D) only if properly controlled experimental conditions can be obtained

3. Of the following factors that affect fire risk, the one that is most adequately reflected in the "unadjusted loss per $1,000 valuation" index is
    (A) fire department operation expenditures
    (B) physical characteristics of the city
    (C) type of structures most prevalent in the city
    (D) total worth of property in the city

4. According to the paragraph, cities that spend larger sums on their fire departments
    (A) tend to have lower fire losses than cities that spend smaller sums on their fire departments
    (B) do not tend to have lower fire losses than cities that spend smaller sums on their fire departments
    (C) tend to have higher fire losses than cities that spend smaller sums on their fire departments
    (D) do not tend to have the same total property valuation as cities that spend smaller sums on their fire departments

Directions: *Questions 5 through 7 relate to the following paragraph and are to be answered in accordance with the paragraph.*

Shafts extending into the top story, except those stair shafts in which the stairs do not continue to the roof, shall be carried through and at least two feet above the roof. Every shaft extending above the roof, except open shafts and elevator shafts, shall be enclosed at the top with a roof of materials having a fire-resistant rating of one hour and a metal skylight covering at least three-quarters of the area of the shaft in the top story. Skylights over stair shafts shall have an area not less than one-tenth the area of the shaft in the top story but shall have an area not less than 15 square feet in area. Any shaft terminating below the top story of a structure and those stair shafts not required to extend through the roof shall have the top enclosed with materials having the same fire-resistant rating as required for the shaft enclosure.

5. The paragraph states that the elevator shafts that extend into the top story
   (A) are not required to have a skylight but are required to extend at least two feet above the roof
   (B) are neither required to have a skylight nor to extend above the roof
   (C) are required to have a skylight covering at least three-quarters of the area of the shaft in the top story and to extend at least two feet above the roof
   (D) are required to have a skylight covering at least three-quarters of the area of the shaft in the top story but are not required to extend above the roof

6. Of the following skylights, the one that meets the requirements of the paragraph is a skylight measuring
   (A) 4 feet by 4 feet over a stair shaft that, on the top story, measures 20 feet by 9 feet
   (B) 4½ feet by 3½ feet over a pipe shaft that, on the top story, measures 5 feet by 4 feet
   (C) 2½ feet by 1½ feet over a dumbwaiter shaft that, on the top story, measures 2½ feet by 2½ feet
   (D) 4 feet by 3 feet over a stair shaft that, on the top story, measures 15 feet by 6 feet

7. Suppose a shaft that does not go to the roof is required to have a three-hour fire-resistant rating. In regard to the material enclosing the top of this shaft, the paragraph
   (A) states that a one-hour fire-resistant rating is required
   (B) states that a three-hour fire-resistant rating is required
   (C) implies that no fire-resistant rating is required
   (D) neither states nor implies anything about the fire-resistant rating

Directions: *Questions 8 through 14 relate to the following paragraphs and are to be answered in accordance with these paragraphs.*

Fire regulations require that every liquefied petroleum gas installation should be provided with the means for shutting off the supply to a building in case of an emergency. The installation of a shut-off valve immediately inside a building, which sometimes is done for the convenience of the user, does not comply with this regulation. An outside shut-off valve just outside the building seems to be the logical solution. However, the possibility of tampering illustrates the danger of such an arrangement. A shut-off valve so located might be placed in a locked box. This has no advantage, however, over a valve provided within the locked cabinet containing the cylinder or an enclosure provided over the top of the cylinder.

Keys can be carried by firefighters or, in an emergency, the lock can be broken. Where no valve is visible, the firefighters should not hesitate to break the lock to the cylinder enclosure. The means for shutting off the gas varies considerably in the numerous types of equipment in use. When the cover to the enclosure has been opened, the gas can be shut off as follows:

Close the tank or cylinder valves to which the supply line is connected. Such valves always turn to the right. If the valve is not provided with a handwheel, an adjustable wrench can be used. If conditions are such that shutting off the supply at once is imperative and this cannot be accomplished as previously mentioned, the tubing that is commonly employed as the supply line can be flattened to the extent of closure by a hammer. If the emergency requires the removal of the cylinder, the supply line should be disconnected and the cylinder removed to a safe location. A tank buried in the ground is safe against fire. When conditions indicate the need to remove a cylinder or tank and this cannot be done due to the severity of exposure, pressures within the container can be kept within control of the safety valve by means of a hose stream played on the surface of the container. The melting of the fuse plug also can be prevented in this way.

8. According to the preceding paragraphs, in an emergency, a firefighter should break the lock of a cylinder enclosure whenever the shut-off valve
    (A) fails to operate
    (B) has no handwheel
    (C) has been tampered with
    (D) cannot be seen

9. According to the preceding paragraphs, shut-off valves for liquefied petroleum gas installations
    (A) always turn to the right
    (B) always turn to the left
    (C) sometimes turn to the right and sometimes turn to the left
    (D) generally are pulled up

10. According to the preceding paragraphs, if a cylinder needs to be moved but cannot be because of the severity of exposure, the pressure can be kept under control by
    (A) opening the shut-off valve
    (B) playing a hose stream on the cylinder
    (C) disconnecting the supply line into the cylinder
    (D) removing the fuse plug

11. The preceding paragraphs state that the supply line should be disconnected when the
    (A) fuse plugs melt
    (B) cylinder is removed to another location
    (C) supply line becomes defective
    (D) cylinder is damaged

12. The preceding paragraphs state that the shut-off valves for liquefied petroleum gas installations are sometimes placed inside buildings
    (A) so that firefighters will be able to find the valves more easily
    (B) because it is more convenient for the occupants
    (C) to hide the valves from public view
    (D) because this makes it easier to keep the valves in good working condition

13. It is suggested in the preceding paragraphs that, during an emergency, the supply line tubing should be flattened to the extent of closure when the
    (A) supply line becomes defective
    (B) shut-off valve cannot be opened
    (C) shut-off valve cannot be closed
    (D) supply line is near a fire

14. According to the preceding paragraphs, fire regulations require that liquefied petroleum gas installations should
    (A) be made in safe places
    (B) be tamperproof
    (C) have shut-off valves
    (D) not exceed a certain size

Directions: Questions 15 through 24 relate to the following paragraphs and are to be answered in accordance with these paragraphs.

Air conditioning systems are complex and are made up of several processes. The circulation of air is produced by fans and ducts; the heating is produced by steam, hot water coils, coal, gas, or oil fire furnaces; the cooling is done by ice or mechanical refrigeration; and the cleaning is done by air washers or filters.

Air-conditioning systems in large buildings generally should be divided into several parts with wholly separate ducts for each part or floor. The ducts are then extended through fire partitions. As a safeguard, whenever ducts pass through fire partitions, automatic fire dampers should be installed in the ducts. Furthermore, the ducts should be lined on the inside with fire-resistant materials. In addition, a manually operated fan shut-off should be installed at a location that is readily accessible under fire conditions.

Most air-conditioning systems recirculate a considerable portion of the air. When this is done, an additional safeguard has to be taken to have the fan arrange to shut down automatically in case of fire. A thermostatic device in the return air duct will operate the shut-off device whenever the temperature of the air coming to the fan becomes excessive. The air filters frequently are coated with oil to help catch dust. Such oil should be of a type that does not ignite readily. Whenever a flammable or toxic refrigerant is employed for air cooling, coils containing such a refrigerant should not be inserted in any air passage.

15. According to the preceding paragraphs, fan shut-offs in the air-conditioning system should be installed
    (A) near the air ducts
    (B) next to fire partitions
    (C) near the fire dampers
    (D) where they can be reached quickly

16. On the basis of the preceding paragraphs, whenever a fire breaks out in a building containing an air-conditioning system that recirculates a portion of the air,
    (A) the fan will shut down automatically
    (B) the air ducts will be opened
    (C) the thermostat will cease to operate
    (D) the fire partitions will open

17. The preceding paragraphs state that, on every floor of a large building in which air-conditioning systems are used, there should be
    (A) an automatic damper
    (B) a thermostatic device
    (C) an air filter
    (D) a separate duct

18. From the preceding paragraphs, the conclusion can be drawn that, in an air-conditioning system, flammable refrigerants
    (A) can be used if certain precautions are observed
    (B) should be used sparingly and only in air passages
    (C) should not be used under any circumstances
    (D) might be more effective than other refrigerants

19. According to the preceding paragraph, the spreading of dust by means of fans in the air-conditioning system is reduced by
    (A) shutting down the fan automatically
    (B) lining the inside of the air duct
    (C) cleaning the circulated air with filters
    (D) coating the air filters with oil

20. According to the preceding paragraphs, the purpose of a thermostatic device is to
    (A) regulate the temperature of the air-conditioning system
    (B) shut off the fan when the temperature of the air rises
    (C) operate the fan when the temperature of the air falls
    (D) assist in the recirculation of the air

21. According to the preceding paragraphs, hot water coils in the air-conditioning system are limited to
    (A) cooling
    (B) heating
    (C) heating and cooling
    (D) heating and cleaning

22. The parts of an air-conditioning system that the preceding paragraphs state should be made of fire-resistant materials are the
    (A) hot water coils
    (B) automatic fire dampers
    (C) air duct linings
    (D) thermostatic devices

23. According to the preceding paragraphs, automatic fire dampers should be installed
    (A) on oil-fired furnaces
    (B) on every floor of a large building
    (C) in ducts passing through fire partitions
    (D) next to the hot water coils

24. On the basis of the preceding paragraphs, the most accurate statement is that the coils containing toxic refrigerants should be
    (A) used only when necessary
    (B) lined with fire-resistant materials
    (C) coated with nonflammable oil
    (D) kept out of any air passage

Directions: Questions 25 through 27 relate to the following paragraph and are to be answered in accordance with this paragraph.

> It shall be unlawful to place, use, or maintain in a condition intended, arranged, or designed for use any gas-fired cooking appliance, laundry stove, heating stove, range, water heater, or combination of such appliances in any room or space used for living or sleeping in any new or existing multiple dwelling unless such room or space has a window opening to the outer air or such gas appliance is vented to the outer air. All automatically operated gas appliances shall be equipped with a device that shuts off automatically the gas supply to the main burners when the pilot light in such appliances is extinguished. A gas range or the cooking portion of a gas appliance incorporating a room heater shall not be deemed an automatically operated gas appliance. However, burners in gas ovens and broilers that can be turned on and off or ignited by nonmanual means shall be equipped with a device that shall shut off automatically the gas supply to those burners when the operation of such nonmanual means fails.

25. According to this paragraph, an automatic shut-off device is not required on a gas
    (A) hot water heater
    (B) laundry dryer
    (C) space heater
    (D) range

26. According to this paragraph, a gas-fired water heater is permitted
    (A) only in kitchens
    (B) only in bathrooms
    (C) only in living rooms
    (D) in any type of room

27. An automatic shut-off device shuts off
    (A) the gas range
    (B) the pilot light
    (C) the gas supply
    (D) all of these

Directions: Questions 28 through 30 relate to the following paragraph and are to be answered in accordance with this paragraph.

> A utility plan is a floor plan that shows the layout of a heating, electrical, plumbing, or other utility system. Utility plans are used primarily by the persons responsible for the utilities, but they are important to the craftsman as well. Most utility installations require the leaving of openings in walls, floors, and roofs for the admission or installation of utility features. The craftsman pouring a concrete foundation wall, for example, must study the utility plans to determine the number, sizes, and locations of the openings he must leave for piping, electrical lines, and the like.

28. Of the following items of information, the one that is *least* likely to be provided by a utility plan is the
    (A) location of the joists and frame members around stairwells
    (B) location of the hot water supply and return piping
    (C) location of light fixtures
    (D) number of openings in the floor for radiators

29. According to the paragraph, of the following, the persons who most likely will have the greatest need for the information included in a utility plan of a building are those who
    (A) maintain and repair the heating system
    (B) clean the premises
    (C) put out the fires
    (D) advertise property for sale

30. According to the paragraph, a repair crew member should find it most helpful to consult a utility plan when information is needed about the
    (A) thickness of all doors in the structure
    (B) number of electrical outlets located throughout the structure
    (C) dimensions of each window in the structure
    (D) length of a roof rafter

Answer Key

1. C	11. B	21. B
2. B	12. B	22. C
3. D	13. C	23. C
4. B	14. C	24. D
5. A	15. D	25. D
6. B	16. A	26. D
7. B	17. D	27. C
8. D	18. A	28. A
9. A	19. D	29. A
10. B	20. B	30. B

Explanatory Answers

1. **(C)** The last sentence states that part of the lack of relationship between the cost of running a fire department and dollar losses in fires is due to inefficiencies in the use of fire department funds.

2. **(B)** The paragraph makes it clear that many factors affect the "unadjusted loss per $1,000" figure. Comparisons to the fire departments of different cities must allow for these other factors.

3. **(D)** The "loss per $1,000" figure is based on the total value of property in the city.

4. **(B)** The paragraph simply states that spending more on the fire department does not guarantee lower total property loss. It does NOT, however, state that the city that spends more has greater losses.

5. **(A)** The first sentence states that all shafts except stair shafts in which the stairs do not continue to the roof must be carried at least two feet above the roof. Elevator shafts must extend those two feet. The second sentence makes the exception that elevator shafts need not have a skylight.

6. **(B)** The 15.75-square-foot skylight over the 20-square-foot pipe shaft covers more than three-quarters of the area of the shaft. All other skylights are inadequate.

7. **(B)** According to the last sentence, a shaft terminating below the top story of a structure must have a top with the same fire-resistant rating as that required for the shaft enclosure (here three hours).

8. **(D)** The third from the last sentence of the first paragraph states that, where no valve is visible, the firefighters should break the lock to the cylinder enclosure.

9. **(A)** The second sentence of the second paragraph states that these tank valves always turn to the right.

10. **(B)** This information is given in the next to last sentence.

11. **(B)** See the fourth sentence of the second paragraph.

12. **(B)** The second sentence states that the shut-off valve placed inside a building for the convenience of the user does not satisfy the emergency shut-off requirements.

13. **(C)** If the shut-off valve cannot be closed, the supply of gas can be shut off by flattening the supply line tubing.

14. **(C)** See the first sentence.

15. **(D)** The last sentence of the second paragraph states that manual fan shut-offs should be installed where they are readily accessible.

16. **(A)** The second sentence of the last paragraph describes a thermostatic device that automatically shuts off recirculation of air in case of fire.

17. **(D)** See the first sentence of the second paragraph.

18. **(A)** The last sentence states that, when flammable refrigerants are used, the coils containing the refrigerant should not be inserted in the air passages. This constitutes a precaution that makes the use of flammable refrigerants acceptable.

19. **(D)** See the third from the last sentence.

20. **(B)** This is detailed in the second sentence of the last paragraph.

21. **(B)** See the second sentence.

22. **(C)** See the next to the last sentence of the second paragraph.

23. **(C)** See the third sentence of the second paragraph.

24. **(D)** Again, this answer is in the last sentence.

25. **(D)** An automatic shut-off device is required on all automatically operated gas appliances; however, a gas range is not considered to be an automatically operated gas appliance.

26. **(D)** A gas-fired water heater is permitted anywhere provided that the space has a window opening to the outer air or the appliance is vented to the outer air.

27. **(C)** The automatic shut-off device shuts off the gas supply to the burners when the pilot light goes out. This prevents the room from filling with gas.

28. **(A)** The utility plan shows the layout of a utility system, not the structure of the building itself.

29. **(A)** The heating system is one of the utility systems, so the people concerned with its maintenance and repair would find it most useful.

30. **(B)** Again, consider the definition of a utility. The electrical system is a utility system. Structural information appears on a utility plan only where it is incidental to the layout of utilities.

REASONING AND JUDGMENT

1. Of the following, the chief advantage of having firefighters under competitive civil service is that
   (A) fewer fires tend to occur
   (B) fire prevention becomes a reality instead of a distant goal
   (C) the efficiency level of the personnel tends to be raised
   (D) provision can then be made for training by the municipality

2. Of the following, it is least likely that fire will be caused by
   (A) arson
   (B) poor building construction
   (C) carelessness
   (D) inadequate supply of water

Directions: *The following three questions, numbered 3 through 5, are to be answered based on the following table. Data for certain categories have been omitted from the table. You are to calculate the missing numbers if needed to answer the questions.*

	Last Year	This Year	Increase
Firefighters	9,326	9,744	
Lieutenants	1,355	1,417	
Captains		433	107
Others	_____	_____	_____
TOTAL	11,469	12,099	

3. The number in the Others group last year was most nearly
   (A) 450
   (B) 475
   (C) 500
   (D) 525

4. The group that had the largest percentage of increase was
   (A) Firefighters
   (B) Lieutenants
   (C) Captains
   (D) Others

5. This year, the ratio between firefighters and all other ranks of the uniformed force was mostly nearly
   (A) 5:1
   (B) 4:1
   (C) 2:1
   (D) 1:1

6. The ideal is that, in big cities, a fire alarm box can be seen from any corner. This is a desirable condition mainly because
    (A) several alarms can be sounded by one person running from one box to another
    (B) little time is lost in sounding an alarm
    (C) an alarm can be sounded from a different box if the nearest one is out of order
    (D) fire apparatus can quickly reach any box

7. Suppose hydrants with a flowing capacity of less than 500 gallons per minute were to be painted red, hydrants with a flowing capacity between 500 and 1,000 gallons per minute were to be painted yellow, and hydrants with a flowing capacity of 1,000 gallons or greater per minute were to be painted green. The principal advantage of such a scheme is that
    (A) fewer fires would occur
    (B) more water would become available at a fire
    (C) citizens would become more acutely aware of the importance of hydrants
    (D) firefighters would save time

8. "At two o'clock in the morning, Mrs. Smart awakened her husband and told him the house was on fire. Mr. Smart dressed hurriedly and ran 17 blocks—past five fire alarm boxes—to the fire station to tell the firefighters that his house was on fire. When Mr. Smart and the firefighters returned, the house had burned down." This is an illustration of the
    (A) need for a plentiful supply of fire alarm boxes
    (B) need for more fire stations
    (C) necessity of preventing fires
    (D) desirability of educating the public

9. "Installation of a modern fire alarm system will mean smaller fires." Of the following, the best justification for this statement is that
    (A) if summoned quickly, firefighters can control a fire before it has a chance to spread
    (B) if the alarm system is modern, firefighters can be given a complete picture of a fire even before they respond
    (C) some fires, such as fires resulting from explosions, assume large proportions in a few seconds
    (D) most industrial establishments depend on more than one method of transmitting fire alarms

10. "The firefighter must not only know how to make the proper knots, he must know how to make them quickly." Of the following, the chief justification for this statement is that
    (A) a firefighter uses many kinds of knots
    (B) a slowly tied knot slips more readily than one tied quickly
    (C) haste makes waste
    (D) in most fire operations, speed is important

11. A survey of dip tank fires in large cleaning establishments shows that the majority of these firms started with an obvious hazard such as cutting and welding torches. This experience indicates most strongly a need for
    (A) a law prohibiting the use of welding torches
    (B) fire safety programs in this industry
    (C) a law requiring welders to be licensed
    (D) eliminating the use of dip tanks

12. To reduce the number of collisions between fire apparatus and private automobiles, a number of suggestions have been made regarding the training and selection of drivers. Of the following courses of action, the one that is *least* likely to lead to the reduction of accidents is to
    (A) select drivers who have the longest driving experience
    (B) require all members of a company to drive fire apparatus
    (C) select drivers who have experience in driving trucks
    (D) select an alternate driver for every piece of fire apparatus

13. "Automobile parking and double-parking on city streets is daily becoming a greater menace to effective firefighting." Of the following effects of automobile parking, the *least* serious is that
    (A) access to the fire building might be obstructed
    (B) ladder and rescue work might be retarded
    (C) traffic congestion might make it difficult for fire apparatus to get through
    (D) greater lengths of hose might be needed in fighting a fire

14. "Sometimes at a fire, a firefighter endures great punishment to reach a desired position that could have been reached more quickly and with less hardship." Of the following, the chief implication of the preceding statement is that
    (A) one should take time off during a fire to completely think through problems presented
    (B) courage is an important asset in a firefighter
    (C) training is an important factor in firefighting
    (D) modern firefighting requires alert firefighters

15. The statement "Municipal inspections should be coordinated" suggests most nearly that
    (A) fire hazards are dangerous
    (B) the detection of fire hazards is connected with other problems found in a municipality
    (C) municipal and fire inspection services are of unequal importance
    (D) municipal inspections are closely tied to noninspectional services

16. "With fireproof schools, it would appear that drills are unnecessary." The main reason for believing this statement to be false is that
    (A) panic sometimes occurs
    (B) fire extinguishers are available in every school
    (C) fire alarms are easily sounded
    (D) children are accustomed to drilling

17. "Flame is of varying heat according to the nature of the substance producing it." This means most nearly that
    (A) some fires are larger than others
    (B) the best measure of the heat produced by a particular substance is its temperature
    (C) there can be no fire without flame or flame without a fire
    (D) the degree of heat evolved by the combustion of different materials is not identical

18. "No firefighter is expected to use a measuring stick to determine the exact amount of hose needed to stretch a line from the entrance of a burning building to the seat of a fire. He should be guided generally by the following rule—one length of hose for every story." This rule assumes most directly some degree of uniformity in
    (A) fire hazards
    (B) building construction
    (C) causes of fires
    (D) window areas

19. "A fire will occur when a flammable material is sufficiently heated in the presence of oxygen." It follows from this statement that a fire will occur whenever there is sufficient
    (A) heat, hydrogen, and fuel
    (B) oxygen, hydrogen, and fuel
    (C) heat, air, and fuel
    (D) gasoline, heat, and fuel

20. Suppose the vast majority of the fires that took place in 1980 in New York City occurred in a particular type of building structure. Before a firefighter could reasonably infer that this particular type of building structure was unduly susceptible to fires in New York City in 1980, he or she would have to know
    (A) the precise incidence, in terms of percent, represented by the somewhat loose phrase "vast majority"
    (B) the frequency of this type of building structure as compared with other types of building structures
    (C) whether 1980 is validly to be taken as the fiscal year or the calendar year
    (D) whether New York City can legitimately be considered as representative of the country as a whole in regard to the statistical frequency of this type of building structure

21. "Every fire is a potential conflagration." Of the following, the most valid inference that can be drawn from this statement is that
    (A) no matter how insignificant a fire might appear to be, a first effort should be to isolate it
    (B) the method of fighting a potential fire must be adapted to the unique circumstances surrounding that fire
    (C) the apparatus and firefighters sent immediately to a fire should be sufficient to handle practically any conflagration that might be found
    (D) the full potentialities of a fire are usually realized

22. Suppose that, for a certain period of time studied, the percentage of telephone alarms that were false alarms was less than the percentage of fire box alarms that were false alarms. Of the following, the one that is most accurate, based solely on the preceding statement, is that during the period studied
    (A) more alarms were transmitted by telephone than by fire box
    (B) more alarms were transmitted by fire box than by telephone
    (C) relatively fewer false alarms were transmitted by telephone than by fire box
    (D) relatively fewer false alarms were transmitted by fire box than by telephone

23. "There are more engine companies than hook and ladder companies. However, to conclude that the number of firefighters assigned to engine companies exceeds the number of firefighters assigned to hook and ladder companies is to make a basic assumption." Of the following, the most accurate statement of the basic assumption referred to in this quotation is that
    (A) the number of hook and ladder companies does not differ greatly from the number of engine companies
    (B) about the same number of firefighters, on the average, are assigned to each type of company
    (C) an engine company, on the average, has fewer firefighters than a hook and ladder company
    (D) the largest engine company is no larger than the largest hook and ladder company

24. If the fire company due to arrive first arrived one minute before the company due second, and the company due third arrived four minutes after the alarm was sent, then the most accurate of the following statements is that
    (A) the second company due arrived three minutes before the third company due
    (B) the third company due arrived three minutes after the first company due
    (C) the second company due arrived two minutes after the alarm was sent
    (D) second company due arrived one minute after the first company due

25. "During June, 25 percent of all fires in a city were in buildings of type A, 40 percent were in buildings of type B, and 15 percent were in buildings of type C." Of the following, the most accurate statement is that the total number of fires during June was
    (A) equal to the sum of the percentages of fires in the three types of buildings, divided by 100
    (B) less than 100 percent
    (C) one-fourth the number of fires in buildings of type B
    (D) four times the number of fires in buildings of type A

26. "One or more public ambulances can be called to a street box by sending on the Morse key therein the following signals in the order given: the preliminary signal 777, the number of the street box, and the number to indicate how many ambulances are wanted." In accordance with these instructions, the proper signal for calling four ambulances to box 423 is
    (A) 4-777-423
    (B) 423-777-4
    (C) 7777-423
    (D) 777-423-4

27. The fire alarm box is an important element of the city's fire protection system. Of the following factors, the one that is of *least* value in helping a citizen send a fire alarm quickly by means of a fire alarm box is that
    (A) the mechanism of a fire alarm box is simple to operate
    (B) fire alarm boxes are placed at very frequent intervals throughout the city
    (C) there are several different types of fire alarm boxes in use
    (D) specific directions for sending an alarm appear on each fire alarm box

28. The number of fire companies in a city is chiefly determined by all of the following conditions except
    (A) the number of hose streams likely to be required to handle such fires as might be expected
    (B) the manpower and capacity of fire department pumping apparatus and of ladder service
    (C) the type of fire alarm system in use
    (D) the accessibility of various parts of the city to fire companies

29. It is desirable for the fire department to have ladders of varying lengths mainly because some
    (A) fires occur at greater distances from the ground than others
    (B) firefighters are more agile than others
    (C) firefighters are taller than others
    (D) fires are harder to extinguish than others

30. The lever on fire alarm boxes for use by citizens should be
    (A) very easy to manipulate
    (B) just a little difficult to manipulate
    (C) very difficult to manipulate
    (D) constructed without regard to ease of manipulation

31. A large fire occurs that you, as a firefighter, are helping to extinguish. An emergency arises and you believe that a certain action should be taken. Your superior officer directs you to do something else that you consider to be undesirable. You should
    (A) take the initiative and follow what you originally thought to be the superior line of action
    (B) think the matter over for a few minutes and weigh the virtues of the two lines of action
    (C) waste no time but refer the problem immediately to another superior officer
    (D) obey orders despite the fact that you disagree

32. Suppose your company is extinguishing a very small fire in a parked automobile. Your commanding officer directs you to perform some act that, as far as you can see, is not going to help in any way to put out the fire. Of the following, the best reason for obeying the order instantly and without question is that
    (A) the fire department is a civil organization
    (B) your officer, after all, has been in the service for a much longer period of time than you have been
    (C) without discipline, the efficiency of your company would be greatly reduced
    (D) the first duty of the commanding officer is to command

33. Assume that you are a firefighter. While you are walking along in a quiet residential neighborhood at about 3 A.M. on a Sunday morning, another pedestrian calls your attention to smoke coming from several windows on the top floor of a three-story apartment house. Of the following, the best action for you to take immediately is to
    (A) race through the house, wake the tenants, and lead them to safety
    (B) run to the nearest fire alarm box, turn in an alarm, and then run back into the house to arouse the tenants
    (C) run to the nearest fire alarm box, turn in an alarm, and stay there to direct the fire apparatus when it arrives
    (D) direct the other pedestrian to the nearest alarm box with directions to stay there after sending the alarm while you go to the apartment from which the smoke is issuing

34. Assume that you are a firefighter, off duty and in uniform in the basement of a department store. A large crowd is present. There are two stairways, 50 feet apart. A person you cannot see screams, "Fire!" You should first
    (A) rush inside and sound the alarm at a fire alarm box
    (B) find the person who shouted "Fire!", ascertain where the fire has occurred, and proceed to extinguish the fire
    (C) jump on top of a nearby counter, order everyone to be quiet and not to move, find out who screamed "Fire!", and reprimand that person publicly
    (D) jump on top of a counter, obtain the attention of the crowd, direct the crowd to walk to the nearest stairway, and announce that there is no immediate danger

35. A number of prominent municipal officials are extremely interested in the operation of the local fire department. These officials frequently arrive at the scene of a fire almost as quickly as the firefighters. It is desirable that the fire department welcome the interest of these municipal officials mainly because of the probability that
    (A) the officials will acquire considerable technical information
    (B) the organization of the fire department will be modified
    (C) fire administration will remain unaffected
    (D) the work and problems of the fire department will receive due recognition

36. The owner of a building at which you helped put out a fire complains bitterly to you that the firefighters broke a number of cellar windows even before setting out to extinguish the fire in the cellar. Of the following, the best action for you to take is to
    (A) question the validity of the data as described by the owner
    (B) request that the owner put the statement in writing
    (C) explain the reason for breaking the windows
    (D) suggest that the owner have the cellar windows replaced with unbreakable glass

37. An elderly man approaches you, as an officer on duty, to complain that the noise of apparatus responding to alarms wakes him in the middle of the night. Of the following, the best way for you to handle this situation is to
    (A) quickly change the subject because the man obviously is a crank
    (B) ask the man for specific instances of apparatus making noise at night
    (C) explain the need for speed in the response of an apparatus
    (D) promise to avoid making unnecessary noise at night

38. A firefighter assigned to deliver a talk before a civic organization first made a detailed outline of his talk. Then, with the help of two members of his company, he prepared several demonstrations and charts. The firefighter's procedure was
    (A) good, chiefly because preparation increases the effectiveness of a talk
    (B) poor, chiefly because a talk should be guided by the questions and comments of the audience and should not follow a predetermined outline
    (C) good, chiefly because the audience will be impressed by the care the firefighter has taken in preparing the talk
    (D) poor, chiefly because the talk will appear rehearsed rather than spontaneous and natural

39. Suppose you were assigned to be in charge of a new headquarters bureau that will have extensive correspondence with the public and very frequent mail contact with the other city departments. It is decided at the beginning that all communications from the headquarters officer are to go out over your signature. Of the following, the most likely result of this procedure is that
    (A) the administrative head of the bureau will spend too much time preparing correspondence
    (B) execution of bureau policy will be unduly delayed
    (C) subordinate officers will tend to avoid responsibility for decisions based on bureau policy
    (D) uniformity of bureau policy as expressed in such communications will tend to be established

40. Suppose you have been asked to answer a letter from a local board of trade requesting certain information. You find that you cannot grant this request. The best way to begin your answering letter is by
    (A) quoting the laws or regulations that forbid the release of this information
    (B) stating that you are sorry the request cannot be granted
    (C) explaining in detail the reasons for your decision
    (D) commending the organization for its service to the community

41. The relationship between the fire department and the press is like a two-way street. The press not only is a medium through which the fire department releases information to the public, the press also
    (A) is interested in the promotion of the department's program
    (B) can teach the department good public relations
    (C) makes the department aware of public opinion
    (D) provides the basis for community cooperation with the department

42. Assume that you have been asked to prepare an answer to a request from a citizen of the city addressed to the chief of the department. If this request cannot be granted, it is most desirable that the answering letter begin by
    (A) telling the person you were glad to receive the request
    (B) listing the laws that make granting the request impossible
    (C) saying that the request cannot be granted
    (D) discussing the problem presented and showing why the fire department cannot be expected to grant the request

43. Assume that you have been requested by the chief of the department to prepare for public distribution a statement dealing with a controversial matter. For you to present the department's point of view in a terse statement, making no reference to any other matter, is in general
    (A) undesirable; you should show all the statistical data you used, how you obtained the data, and how you arrived at the conclusions presented
    (B) desirable; people will not read long statements
    (C) undesirable; the statement should be developed from ideas and facts familiar to most readers
    (D) desirable; the department's viewpoint should be made known in all controversial matters

44. The most important function of good public relations in a fire department should be in
    (A) training personnel
    (B) developing an understanding of fire dangers on the part of the public
    (C) enacting new fire laws
    (D) recruiting technically trained firefighters

45. The primary purpose of the public relations program of an administrative organization should be to develop mutual understanding between
    (A) the public and those who benefit from organized service
    (B) the public and the organization
    (C) the organization and its affiliate organizations
    (D) the personnel of the organization and the management

46. You arrive in your office at 11 AM, having been on an inspection tour since 8 AM. A man has been waiting in your office for two hours. He is abusive because of his long wait, and he accuses you of sleeping off a hangover at the taxpayers' expense. You should
    (A) say that you have been working all morning and let him sit in the outer office a little longer until he cools off
    (B) tell him you are too busy to see him and make an appointment for later in the day
    (C) ignore his comments, courteously find out what his business is, and take care of him in a perfunctory manner
    (D) explain briefly that your duties sometimes take you out of your office and that an appointment would have prevented the inconvenience

47. During the routine inspection of a building, a citizen tells you very unfavorable personal comments concerning several top officials of the department. Of the following, it usually would be best for you to
    (A) try to change the subject as soon as possible
    (B) attempt to convince the citizen that he is in error
    (C) advise the citizen that your opinion might be like his but that you can't discuss it
    (D) tell the citizen that it would be more proper for him to put his comments into writing

48. "The fire officer should, as far as operations permit, answer any reasonable questions asked by the occupant of the premises involved in a fire." Of the following, however, the subject about which he should be most guarded in his comments is the
    (A) time of the arrival of companies
    (B) methods used to control and extinguish the fire
    (C) specific cause of the fire
    (D) probable extent of the damage

49. The public is most likely to judge personnel largely on the basis of their
    (A) experience
    (B) training and education
    (C) civic-mindedness
    (D) manner and appearance while on duty

50. When employees consistently engage in poor public relations practices, of the following, the cause is most often
    (A) disobedience of orders
    (B) lack of emotional control
    (C) bullheadedness
    (D) poor supervision

51. A citizen's support of a fire department program can best be enlisted by
    (A) minimizing the cost to the community
    (B) telling him how backward the community is in its practices and why such a situation exists
    (C) telling him that it is his civic duty to do all that he can to support the program
    (D) informing him how the program will benefit him and his family

52. The term "public relations" has been defined as the aggregate of every effort made to create and maintain good will and to prevent the growth of ill will. This concept assumes particular importance with regard to public agencies because
    (A) public relations become satisfactory in inverse ratio to the number of personnel employed in the agency
    (B) legislators might react unfavorably to the public agency no matter how its public relations policy is developed
    (C) they are much more dependent on public good will than commercial organizations are
    (D) they are tax-supported and depend on the active and intelligent support of an informed public

53. Of the following, the proper attitude for an administrative officer to adopt toward complaints from the public is that he or she should
    (A) not only accept complaints but establish a regular procedure whereby they can be handled
    (B) avoid encouraging correspondence with the public on the subject of complaints
    (C) remember that it is his or her duty to get a job done, not to act as a public relations officer
    (D) recognize that complaints are rarely the basis for significant administrative action

54. Of the following types of individuals who pull false alarms, reducing the number of false alarms by education alone is most effective for
    (A) mental patients
    (B) drunks
    (C) children of school age
    (D) former firefighters

55. Excluding direct enforcement measures, the most promising approach to relieve the municipal false alarm problem seems to be concerned with
    (A) better location and distribution of alarm boxes
    (B) replacement of telegraph systems with telephone systems
    (C) improved electrical circuitry for verification
    (D) public relations and education

56. An aspect of public relations that has been severely neglected by government agencies, but that has been stressed in business and industry, is
  (A) use of radio
  (B) public speaking
  (C) advertising
  (D) training employees in personal contacts with the public

57. In dealing with the public, it is helpful to know that most people generally are more willing to do something when they
  (A) are not responsible
  (B) understand the reasons
  (C) will be given a little assistance
  (D) must learn a new skill

58. A study shows that false alarms occur mostly between noon and 1 PM and between 3 and 10 PM. The most likely explanation of these results is that many false alarms are sent by
  (A) school children
  (B) drunks
  (C) mental patients
  (D) arsonists

59. While visiting the lounge of a hotel, a firefighter discovers a fire that apparently has been burning for some time and is rapidly spreading. Of the following, the first action for him to take is to
  (A) find the nearest fire extinguisher and attempt to put out the fire
  (B) notify the desk clerk of the fire and send an alarm from the nearest street alarm box
  (C) send an alarm from the nearest street alarm box only
  (D) run throughout the hotel and warn all occupants to evacuate the building

60. When a fire occurs in the city of Merritt, the alarm transmitted over the fire alarm telegraph system sounds in every firehouse in Merritt. Of the following, the best justification for this practice is that
  (A) some companies are more efficient than others
  (B) to listen to a large number of alarms keeps the firefighters alert
  (C) for maximum effective disposition of fire apparatus, the staff at the central headquarters should be fully aware of the entire situation at any moment
  (D) although certain companies are responding to one alarm, other companies must be prepared to respond to new alarms in the same or nearby districts

61. Local firefighters have been called to the assistance of airport firefighters at the site of a major airplane crash. The primary concern of the firefighters must be
  (A) extinguishment of the fire
  (B) protection of life
  (C) protection of exposures
  (D) overhauling

62. Your fire company has just responded to a fire in the hold of a ship docked at a pier. Prior to your arrival, the ship's crew began to take certain actions to control the fire. Which of their following recent actions would prove *least* effective in controlling the fire?
    (A) They sealed the hatch covers on the hold with waterproof tape.
    (B) They shut down the ventilators from the hold and sealed them.
    (C) They flooded the hold with carbon dioxide.
    (D) They placed thermostats throughout the vessel to detect any increase in temperature.

63. Experience in fighting fires in high-rise structures that don't have sprinklers has created a new body of knowledge for firefighters. Which of the following statements is most correct with regard to lessons learned from such fires?
    (A) Central air conditioning, if allowed to continue to supply fresh air in the vicinity of the fire, will cool the fire down.
    (B) New lightweight construction helps restrict vertical extension of the fire to higher floors.
    (C) Extreme temperatures have no significant impact on a firefighter's effective work time.
    (D) Synthetic materials used in furnishings increase the threat to life and safety.

64. Your fire company responds to a fire involving drums of hazardous waste materials. Of the following actions, which would be the *least* correct for you to take?
    (A) Treat empty drums carefully because they might be more hazardous than full drums.
    (B) Avoid using water on a fuming substance until you have identified the substance.
    (C) Smell or touch the substance involved to get an indication of just what is burning.
    (D) Treat small leaks in small containers very carefully.

65. Your ladder company is operating at a serious fire in an old loft building. During operations, you become aware of a large accumulation of water on the third floor. The depth of the water is greater in the center of the room than on the sides. In addition, large slabs of plaster are falling from the walls. It would be most reasonable for you to
    (A) continue with your assigned firefighting duties. Large amounts of water always are necessary to extinguish a large fire.
    (B) take care to avoid being hit by the falling plaster, a normal occurrence in building fires
    (C) skirt the deep water in the middle of the room. Old buildings generally sag in the middle.
    (D) give this information to your superior officer so he or she can assess the possibility of building collapse

66. As a firefighter, you are summoned to a hazardous waste emergency involving radioactive isotopes. Of the following, the factor *least* likely to present a health hazard is
    (A) atmospheric conditions
    (B) length of exposure
    (C) nature of the radiation
    (D) intensity of the radiation

67. Your fire company responds to the site of a serious auto accident and finds the driver unconscious in the car. The victim's position severely limits your survey of the injuries involved. The vehicle itself, however, might give some clues. Of the following, which would be the *least* correct assumption?
(A) A cracked windshield or rearview mirror might indicate injury to the head and spine.
(B) A damaged steering wheel or column might indicate chest, head, or abdominal injury.
(C) Dents on the dashboard might indicate leg, knee, or pelvis trauma.
(D) Because the driver's seat belt is still in position, injuries to spine or abdomen are not likely.

68. The firefighters in a ladder company are required to search for victims in apartments above the site of a store fire. Of the following, the *least* reasonable procedure is to
(A) have a plan and move toward windows and secondary means of egress
(B) in each bedroom, flip a mattress into a "U" position to indicate that the area has been searched
(C) pause from time to time to listen for crying, moaning, or coughing
(D) have the firefighters, after completing their first search for victims, repeat their search a second time to make sure all victims have been located

69. In the case of an early morning major explosion in which a number of blocks are threatened with being engulfed by a fire, the structure that should demand the *least* attention from firefighters is
(A) a hotel
(B) a theater
(C) a single-family dwelling
(D) an apartment house

70. The fire department has criticized the management of several hotels for failure to call the fire department promptly when fires are discovered. The most probable reason for this delay by the management is that
(A) fire insurance rates are affected by the number of fires reported
(B) most fires are extinguished by the hotel's staff before the fire department arrives
(C) hotel guests frequently report fires erroneously
(D) it is feared that hotel guests will be alarmed by the arrival of fire apparatus

Answer Key

1. C	11. B	21. A
2. D	12. B	22. C
3. A	13. D	23. B
4. C	14. C	24. D
5. B	15. B	25. D
6. B	16. A	26. D
7. D	17. D	27. C
8. D	18. B	28. C
9. A	19. C	29. A
10. D	20. B	30. B

31. D	41. C	51. D	61. B
32. C	42. C	52. D	62. D
33. D	43. C	53. A	63. D
34. D	44. B	54. C	64. C
35. D	45. B	55. D	65. D
36. C	46. D	56. D	66. A
37. C	47. A	57. B	67. D
38. A	48. C	58. A	68. D
39. D	49. D	59. B	69. B
40. B	50. D	60. D	70. D

Explanatory Answers

1. **(C)** The rationale behind the civil service testing program is that the person with the higher score will become a better firefighter. If this is correct, then a firefighting force made up of high scorers on civil service exams will be more efficient than one made up of people appointed on the basis of friendships or political connections.

2. **(D)** A fire might get out of control because of an inadequate supply of water, but the lack of water could not be its cause.

COMPLETED TABLE
(QUESTIONS 3 TO 5)

	Last Year	This Year	Increase
Firefighters	9326	9744	418
Lieutenants	1355	1417	62
Captains	326	433	107
Others	462	505	43
TOTAL	11469	12099	630

3. **(A)** To calculate the number of people in the Others group last year, you must first find the number of Captains last year. Because the increase from last year to this year among Captains was 107, subtract 107 from 433 to find out the number of Captains last year. By adding the number of Firefighters, Lieutenants, and Captains and then subtracting that sum from last year's Total, we find that the number of Others last year was 462. 462 is closest to 450, so the answer to the question is (A).

4. **(C)** To determine percent of change, find the difference between the numbers and divide that difference by the original number: $107 \div 326 = 33\%$. Try the others and you will find that the Captains had the greatest percentage of increase by far.

5. **(B)** There are 9,744 Firefighters and 2,355 people of all other ranks. This constitutes a ratio of approximately 4:1.

6. **(B)** In firefighting, time is crucial. If a fire alarm box can be found at once, no time is lost in sounding the alarm.

7. **(D)** Obviously, the more information available to firefighters, the less time they need to spend finding things out for themselves and the more quickly they can attack the fire.

8. **(D)** If Mr. Smart had known how to use a fire alarm box, he might have alerted the firefighters far more quickly, and his house might have been saved. The public must be educated to use all resources.

9. **(A)** Presumably, a modern fire alarm system enables alarms to be transmitted and received most rapidly and accurately. The more quickly the firefighters can respond, the more quickly they can control the fire and the smaller it will be.

10. **(D)** It is impossible to overemphasize the need for speed in all aspects of firefighting.

11. **(B)** Obviously, the number of industrial fires can be reduced by proper fire safety education of workers in fire-prone industries.

12. **(B)** Experience is the best teacher. If a few firefighters drive frequently, they will become experts. If firefighters rotate the driving responsibility, each will drive less often, and no one will have the opportunity to develop expertise.

13. **(D)** The need for greater lengths of hose is less serious than problems in reaching the site of the fire or in gaining access to the building.

14. **(C)** A better-trained firefighter is more likely to recognize the best way to approach a problem.

15. **(B)** If fire inspections are made in conjunction with other safety inspections, remedial actions and follow-up also can be coordinated for efficiency and guarantees of compliance.

16. **(A)** Fireproof schools do not have fireproof contents. In the case of a fire, evacuation of students is a must. Fire drills are necessary so that teachers and students become familiar with procedures and will not panic in the case of a fire.

17. **(D)** The heat produced by all fires is not alike.

18. **(B)** If one length of hose is required for each story of a building, it must be assumed that the height of a story is similar in most buildings.

19. **(C)** Heat always is necessary for fire. Flammable material is the fuel; oxygen is the component of air that promotes combustion.

20. **(B)** If most of the buildings are of a particular construction, it stands to reason that most of the fires would occur in that kind of building. The answer to this question is entirely based on the frequency of that type of building as compared to all others.

21. **(A)** In other words, even a tiny fire can grow. Take all fires seriously.

22. **(C)** This is a combination reading-comprehension and reasoning question. The correct answer is simply a restatement of the first sentence.

23. **(B)** If the same number of firefighters are assigned to engine companies as to hook and ladder companies, and if there are more engine companies, then there would be more firefighters assigned to engine companies.

24. **(D)** Because the first company arrived one minute before the second, the second company arrived one minute after the first. This is the only certain answer.

25. **(D)** If 25 percent of all fires were in buildings of Type A, the total number of fires was four times the number of fires in Type A buildings.

26. **(D)** The preliminary signal is the first signal (here 777) followed by box number 423 and the four ambulances wanted.

27. **(C)** Quite to the contrary, if directions for using alarm boxes vary from one to the next, the sending of alarms will be delayed while citizens study the directions. It is important for the directions to be uniform throughout the city.

28. **(C)** The type of alarm system has little bearing on the number of fire companies. What is important is that there should be sufficient manpower and sufficient equipment to reach all parts of the city efficiently.

29. **(A)** Longer ladders are more difficult to maneuver, so it is good to have short ladders for low fires. Long ladders, however, must be available to fight fires on higher floors.

30. **(B)** The logic of this answer is that slight difficulty will discourage accidental false alarms.

31. **(D)** There would be chaos if each firefighter were to create his or her own plan of action and follow it. You MUST obey orders. Your superior officer has had greater experience in fighting fires. At the station house after the fire, you might discuss the orders with your superior officer and learn about the reasoning behind them.

32. **(C)** Teamwork is vital in firefighting. Discipline coordinates teamwork.

33. **(D)** Sending in the alarm is the first priority, but the other pedestrian is perfectly qualified to do this chore. Assign the pedestrian to send the alarm and direct the firefighters while you, as a trained firefighter, begin to rouse and rescue.

34. **(D)** The fire is not visible and is not yet a clear and present danger. Of far greater concern is panic and stampede. Deal with the crowd first and, if possible, direct someone to turn in an alarm at the same time.

35. **(D)** As long as municipal officials do not get in the way of firefighters, their interest and appreciation should be most welcome. Recognition is good for morale, and appreciation of the firefighters' work can do no harm at budget time.

36. **(C)** The best course in dealing with irate citizens is patient explanation.

37. **(C)** Again, explanation is the best policy.

38. **(A)** Obviously, preparation makes for an organized, useful talk. Although the audience might be impressed with the firefighter's careful presentation, impressing the public must not be the goal of the firefighter.

39. **(D)** If one person oversees and signs all correspondence, that person has control over the content of the correspondence and can ensure consistency of policy and tactfulness of expression.

40. **(B)** This is a matter of tact and public relations. Begin with your regrets and then explain the reasons for your refusal.

41. **(C)** A free press reports what it hears. The press reflects public opinion; the fire department must listen.

42. **(C)** A business letter, which is what this is, should come straight to the point and then explain in greater detail.

43. **(C)** You must not appear to be arbitrary in dealing with controversial matters with which the public is genuinely concerned. Explain in a way that is clear and reasonable.

44. **(B)** Public relations entails relating to the public. Of the choices, only explaining fire dangers involves dealing with the public.

45. **(B)** The public relations program of an organization has to do with the relations between that organization and the public.

46. **(D)** There is no point in further antagonizing the man. On the other hand, by all means, let him know that you were involved in fire department business. Suggesting that he might have made an appointment is constructive criticism.

47. **(A)** This might be a hard spot to get out of. No professional should discuss his or her superiors with the public. Do your best to find another topic of conversation. Avoid expressing any opinion.

48. **(C)** The firefighter is not qualified to determine the cause of a fire. That is the role of investigators. It is best to refer questions about the cause of a fire to the experts.

49. **(D)** The public bases judgments first on results (not offered here as an answer choice) and then on appearances.

50. **(D)** Occasional poor public relations practices might stem from lack of self-control or disobedience of orders. If this is a consistent problem, it more likely stems from poor supervision and lack of instruction in more desirable practices.

51. **(D)** Self-interest is a powerful stimulus.

52. **(D)** By definition, public agencies are supported by the public in terms of finances and cooperation. Good will is vital.

53. **(A)** A regular procedure for handling complaints leads to quick disposition of complaints and might even minimize their number. A public that knows its concerns are taken seriously is less likely to complain.

54. **(C)** It is impossible to change the behavior of mentally ill persons or drunks through education alone. If former firefighters are turning in false alarms, they have severe psychological problems; they already are fully aware of the dangers of false alarms. Schoolchildren, on the other hand, can be directed away from undesirable behavior through education and explanation.

55. **(D)** The public must be made aware of the dangers of false alarms and the expense to the taxpayers.

56. **(D)** Government agencies tend to neglect employee training in nonperformance areas. Some training in dealing with the public might lead to a better image and greater cooperation for government agencies.

57. **(B)** People always are more willing to cooperate with requests when they understand the reasons. Education is the key.

58. **(A)** From noon to 1 PM is the lunch hour; from 3 PM to 10 PM, school is out. The culprits are schoolchildren.

59. **(B)** The alarm must be sounded immediately for a well-developed fire. The desk clerk is best equipped to notify hotel occupants of the fire and should do that while the firefighter sends in the alarm. The firefighter should then return to assist with evacuation.

60. **(D)** All fire companies must be aware of the location and involvement of all other companies at all times. This way, they know which companies are available to respond to subsequent alarms, and the entire city is certain to be protected at all times.

61. **(B)** The main objective of all firefighting at all times is to save the lives of civilians and other firefighters.

62. **(D)** Thermostats, strategically placed long before a fire, might have given advance warning and might have been useful in preventing the development of the fire. After the ship is actually on fire, however, placing thermostats is akin to closing the barn door after the horses have escaped. All the other actions should be useful in controlling the fire.

63. **(D)** Synthetic materials produce highly toxic products of combustion that seriously increase the threat to life and safety. All the other statements are untrue. Introduction of fresh air contributes to the intensity of a fire. Fire spreads upward rapidly in the new lightweight high-rise construction. At extreme temperatures, a firefighter's effective work time is approximately five minutes.

64. **(C)** Under no circumstances should you touch, smell, or taste any suspected hazardous substance nor any unknown substance involved in a fire. All the other actions would be correct.

65. **(D)** Collapse of the building may be imminent. The weight of water causes beams to sag in the middle. This, in turn, pulls them from their supports on the sides of the building. The large slabs of falling plaster are a further indication of the shifting of floors and walls. The situation in the old loft building is by no means normal or routine.

66. **(A)** Dangers of radiation come from length of exposure, the nature of the radiation, and the intensity of the radiation. Atmospheric conditions have little influence on the hazards from radiation exposure.

67. **(D)** Although the use of seat belts does save lives and can limit many injuries, it is not a 100-percent effective measure. Use of seat belts, especially without the shoulder harness in place, does not always prevent injury to the abdomen, spine, and pelvis. The other assumptions all deserve consideration.

68. **(D)** A second search is very much in order, but it should be made by a different group of firefighters. If the first group has not found victims, it makes sense to send another group that might have different, more effective methods. The other procedures are all proper and helpful.

69. **(B)** In the early morning, a theater is unlikely to be occupied. Because lifesaving is the paramount concern, an unoccupied structure should not divert attention from buildings where lives are in danger.

70. **(D)** Your knowledge of human nature should tell you that the hotel's management is concerned with public relations. Fear of panic among guests and of publicity that the hotel is fire-prone can combine to discourage hotel management from reporting fires promptly.

SPATIAL ORIENTATION

Spatial orientation questions measure your ability to keep a clear idea of where you are in relation to the space in which you happen to be. The ability to orient yourself in space is very important to a firefighter who must reach a fire in a minimum amount of time and in the safest and most effective manner. Spatial orientation also is vital to firefighters in a smoky environment. Knowing the direction in which you are facing after a number of turns with no visual clues might be a skill that will save your life.

Your firefighting training will give you clues and instructions to orient yourself in space. You will have lots of practice before you are put into a position in which you must rely on this skill. The spatial orientation questions on the exam are meant to test your native aptitude in this regard and to measure how carefully you read and how logically you follow through.

Firefighter exam spatial orientation questions tend to emphasize either where you are in a diagram or how to go from one spot to another on the diagram or map. Because these are not memorization questions, you probably will be allowed to use your pencil to write on the diagrams or maps as a way of testing your answer choices. When using your pencil to write on a diagram or map, be sure to write lightly. Erase any of your jottings that do not work out or no longer are needed. If several questions are based on the same diagram or map and you have made pencil markings for them, the diagram can get quite confusing with markings from prior questions. Hence, if there is another question to be answered on the basis of the same diagram, you should erase your markings as soon as you are done with them.

Many diagrams or maps use symbols, also called legends. Look at the whole page to see if there is a key to the legend. A dotted line might indicate movement. An arrow might indicate direction of movement or of permitted movement. An important feature of many diagrams and maps is the symbol indicating the directions of north, south, east, and west.

Questions often are based on phrases such as "turn left," "to the right," or "to the left of the rear entrance." The test-maker often approaches a diagram or map from the side or from the top, however, so that "left" and "right" do not correspond to where you are sitting in relation to the diagram. Just turn the test booklet sideways or upside-down to view the map or diagram from the standpoint of the question. Turn it so that "left" or "right" on the diagram or map is in the same direction as your left or right hand.

Occasionally, a question will give you information in verbal form and will ask you to relate the information to a map or diagram. The question might say that a fire engine went north two blocks, for example, then turned east for two blocks, turned south for one block, and then went east one more block. The answer choices might consist of just lines showing the directions of movement. The correct line will go up, go left, come down halfway, and then go left again. In a question of this sort, you might find it helpful to make a diagram of your own in the margin of the test booklet. If you diagram the problem yourself, you are less likely to be confused by wrong answers.

Directions: Answer question 1 on the basis of the following street map and information. Circle the letter of your answer choice.

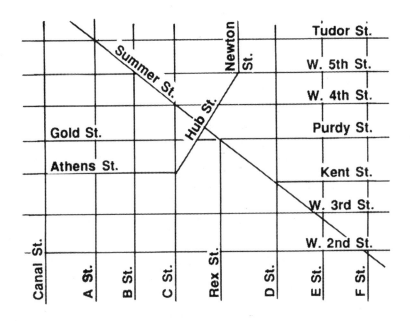

This portion of a city street map shows an area that is divided into four fire control sectors as follows:

–Sector Adam: Bounded by Tudor Street, Newton Street, Hub Street, Athens Street, and Canal Street

–Sector Boy: Bounded by Tudor Street, F Street, West 4th Street, Hub Street, and Newton Street

–Sector Charles: Bounded by West 4th Street, F Street, West 2nd Street, C Street, and Hub Street

–Sector David: Bounded by Athens Street, C Street, West 2nd Street, and Canal Street

1. There is a fire in the block bounded by West 4th Street, Summer Street, and Hub Street. The fire is in sector
   (A) David
   (B) Charles
   (C) Boy
   (D) Adam

Directions: Answer question 2 on the basis of the following passage and diagram.

As indicated by arrows on the following street map, Adams and River Streets are one-way going north. Main is one-way going south, and Market is one-way going northwest. Oak and Ash are one-way streets going east, and Elm is one-way going west.

2. A fire engine heading north on River Street between Ash and Elm Streets receives a call to proceed to the intersection of Adams and Oak. To travel the shortest distance and not break any traffic regulations, the engine should turn
   (A) left on Elm and right on Market
   (B) left on Market and proceed directly to Oak
   (C) left on Elm and right on Adams
   (D) left on Oak and proceed directly to Adams

Directions: *Answer questions 3 through 5 on the basis of the following map.*

The flow of traffic is indicated by the arrows. You must follow the flow of the traffic.

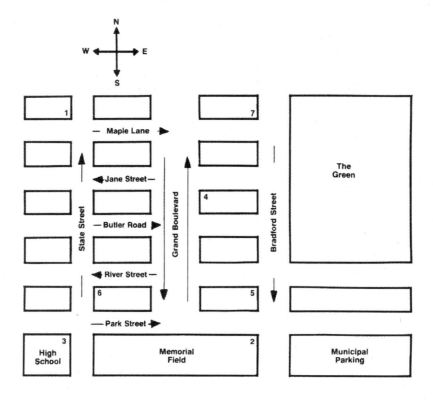

3. If you are located at point (1) and travel east one block, turn right and travel four blocks, and then turn left and travel one block, you will be closest to point
   (A) 6
   (B) 5
   (C) 3
   (D) 2

4. You have responded to an alarm sent from the box at the corner of Butler Road and Bradford Street, and you learn that the fire is at Butler Road and Grand Boulevard. Which is the best route to take to the scene?
   (A) Go one block south on Bradford Street, go one block west on River Street, and then go one block north on Grand Boulevard.
   (B) Go one block south on Bradford Street, go two blocks west on River Street, go one block north on State Street, and then go one block east on Butler Road.
   (C) Go one block north on Bradford Street, go one block west on Jane Street, and then go one block south on Grand Boulevard.
   (D) Go two blocks south on Bradford Street, go one block west on Park Street, and then go two blocks north on Grand Boulevard.

5. You have just checked out and dismissed a false alarm at the high school when you receive word of a trash can fire on The Green near the corner of Maple Lane and Bradford Street. The best route to take would be to
   (A) go two blocks east on Park Street and then go four blocks north on Bradford Street
   (B) go two blocks north on State Street, go two blocks east on Butler Road, and then go two blocks north on Bradford Street
   (C) go four blocks north on State Street and then go two blocks east on Maple Lane
   (D) go one block east on Park Street, go two blocks north on Grand Boulevard, go one block east on Butler Road, and then go two blocks north on Bradford Street

Directions: Answer questions 6 through 12 on the basis of the following diagram.

6. The parade down Main Street has cleared Clark Boulevard and is crossing Hiram Road when an alarm is received of a fire at the Post Office. The best hydrant for the firefighters to attach their first hose to should be the hydrant
   (A) at the corner of Barton Street and Main Street
   (B) mid-block on the northeast side of Main Street
   (C) mid-block on Smith Street between Hiram Road and Barton Street
   (D) at the corner of Hiram Road and Smith Street

7. The best route for the firefighters to take to the Post Office fire in question 6 is
   (A) straight down Barton Street
   (B) north on Bell Road, right on Clark Boulevard, and then south on Smith Street
   (C) north on Bell Road, right onto Hiram Road, and right again onto Main Street
   (D) east on Barton Street, north onto Charles Street, right onto Hiram Road, and right again onto Smith Street

8. At 5 PM, word is received at the firehouse of a fire at the YMCA. The best route to this fire is
   (A) east one block on Barton Street and then left into Charles Street
   (B) north on Bell Road to Clark Boulevard and then right onto Clark Boulevard to Charles Street
   (C) east on Barton Street, left onto Smith Street, and then west onto Clark Boulevard
   (D) north on Bell Road to Hiram Road, east onto Hiram Road, and then north onto Charles Street

9. While the fire at the YMCA is raging, a strong wind is blowing from the southeast. Firefighters should train their hoses to avoid having the fire spread to the
   (A) Plaza Theater
   (B) Parking Garage
   (C) Elegant Apartments
   (D) Textile Factory

10. A fire alarm sent from the alarm box on Main Street between Clark Boulevard and Hiram Road is *least* likely to be reporting a fire
    (A) in an Office Building
    (B) at the Pipe Factory
    (C) at the Elegant Apartments
    (D) at the Catholic Church

11. On a hot summer evening, a firefighter is sent to make a routine check to ensure that no fire hydrants have been opened to reduce water pressure in the area. The firefighter leaves the firehouse by way of the Barton Street entrance and heads east. He takes the second left turn, proceeds one block, and makes another left, goes one block, and makes three successive right turns. At the next intersection, he turns left, then makes his first right, and stops the car. The firefighter is now
    (A) back at the firehouse
    (B) at the corner of Hiram Road and Main Street facing southeast on Main Street
    (C) at the corner of Charles Street and Clark Boulevard facing east on Clark Boulevard
    (D) on Bell Road at the side entrance of the Embassy Hotel

12. The most poorly protected block in terms of a pedestrian's ease in reporting a fire is the block bounded by
    (A) Charles Street, Main Street, and Barton Street
    (B) Hiram Road, Main Street, Barton Street, and Smith Street
    (C) Clark Boulevard, Charles Street, Hiram Road, and Smith Street
    (D) Hiram Road, Bell Road, Barton Street, and Charles Street

Directions: *Answer questions 13 through 16 on the basis of the following floor plan.*

Central Hotel

Central Hotel is an elegant, old, five-story structure. The plan of each residential floor, with four floors above the lobby, is identical. ᛦ represents a standpipe connection.

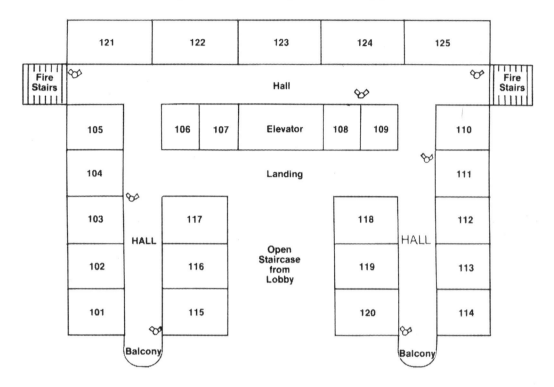

13. In the event of a fire in the hotel lobby, the best means by which to evacuate the tenant in room 124 is
    (A) via the elevator
    (B) down the fire stairs next to room 125
    (C) across the hall and down the wide open staircase
    (D) across the hall and down the corridor to the balcony

14. In the same lobby fire, the tenant in room 217 should be evacuated by
    (A) running down the hall to the balcony
    (B) taking the elevator
    (C) whipping around the corner and down the open staircase
    (D) using the fire stairs between rooms 105 and 121

15. In the case of a fire in room 406, firefighters should
    (A) take the elevator to the roof, enter the fifth floor from the fire stairs, and connect to the standpipe at room 421
    (B) run up the open staircase and connect hoses to the standpipe between rooms 403 and 404
    (C) enter the building by way of a ladder placed in front and up to the balcony, then connect to the standpipe at rooms 403 and 404
    (D) run up the fire stairs and connect to the standpipe at room 421

16. Aside from the danger to the tenant of the involved room, the greatest danger from a fire in room 309 is to the tenants of
    (A) rooms 310 and 318
    (B) rooms 209 and 409
    (C) rooms 308 and 409
    (D) rooms 308 and 209

Answer Key

1. D
2. A
3. D
4. A
5. C
6. C
7. B
8. A
9. D
10. A
11. C
12. B
13. B
14. A
15. D
16. C

Explanatory Answers

1. **(D)** Sector Adam is bounded by heavy black lines. The block with the fire is marked with an X.

2. **(A)** The route is marked on the diagram. Choice (C) would be slightly longer. Choices (B) and (D) are impossible from the given starting point.

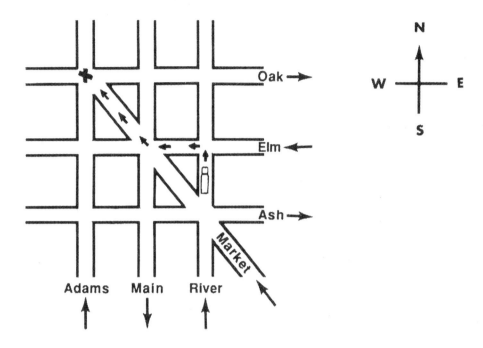

3. **(D)** From point (1), east one block is right on Maple Lane to Grand Boulevard, right four blocks is south on Grand Boulevard to Park Street, left is east onto Park Street, and one block brings us closest to point (2).

4. **(A)** Bradford Road is one-way southbound, so immediately eliminate choice (C). Park Street is one-way eastbound, so eliminate choice (D). Choice (B) is possible but is longer than choice (A) to no advantage.

5. **(C)** All other choices send you north on Bradford Street.

6. **(C)** With a parade coming down Main Street, it would be difficult to maneuver fire trucks and hoses on either side of the parade route. The hydrant at choice (C) is closer than the one at (D).

7. **(B)** Certainly, the route straight down Barton Street is the most direct but remember the parade. Under the special circumstances of the day, the best route is to entirely skirt the parade. The parade has cleared Clark Boulevard, and the detour around the parade route is not really far.

8. **(A)** Look at the time of day. Clark Boulevard is clogged with workers on their way home and should be avoided if at all possible. Choice (D) is acceptable but (A) is faster.

9. **(D)** Wind from the southeast is blowing toward the northwest. The Textile Factory is directly northwest from the YMCA.

10. **(A)** The fire box on Main Street between Clark Boulevard and Hiram Road is between the Ice Cream Shoppe and the Central Bank. Direct vision from this box to the office buildings is completely blocked by the Elegant Apartments. A person at this location is unlikely to be aware of a fire at either office building.

11. **(C)** The firefighter headed east on Barton and made the second left onto Main Street. He went one block and then turned left (west) onto Hiram Road. After one block, he turned right (north) onto Bell Road, right again (east) onto Clark Boulevard, and right once more (southeast) onto Main Street. At the next intersection, he turned left (north) onto Charles Street and then right, placing him on Clark Boulevard facing east. The arrows mark his route:

12. **(B)** There is no fire alarm box on this block. It is quite far from the nearest box on the next block, and it is not close to the firehouse. The block in choice (A) also has no box, but it is closer to the nearest box and to the firehouse.

13. **(B)** Fire, heat, and smoke will rise rapidly up the open staircase from the lobby and will fill the halls of the floors above. Tenants being evacuated should travel the shortest possible distances in the halls. The fire stairs are close and are safe because they are enclosed. Elevators are never safe in a fire. Further, because the fire is in the lobby, patrons should be kept away from the lobby.

14. **(A)** The reasoning is the same as in question 13. This tenant will reach fresh air and safety most quickly and with least exposure to smoke by running to the balcony for rescue from outside.

15. **(D)** As previously mentioned, the elevator and open staircase present hazards to all people—firefighters as well as civilians. The safely enclosed fire stairs are a much quicker and safer route by which to carry hoses to the upper floor.

16. **(C)** Room 308 shares a common wall with room 309 and is clearly in jeopardy. Common walls tend to be flimsier than floors and ceilings. Because heat and smoke rise, the danger is greater to the room above room 309 than to the room below, though it too should be evacuated.

OBSERVATION AND MEMORY

Firefighters do not work alone. Whatever it is they are doing—fire inspections, equipment maintenance, firefighting—they always do their work in pairs, groups, or teams. Much of a firefighter's work is done under direction and supervision. It is important, however, that the firefighter keep alert at all times and not rely solely on the observations and judgments of others.

Fire inspections provide the ideal opportunity for firefighters to learn neighborhood and building layouts, locations of firefighting aids, and possible hazardous situations. It is vital for a firefighter out on inspection to observe everything about the premises. It is to be hoped that observation of problem situations during inspections will lead to corrective and preventive measures so that fires never occur at the premises.

Fires do occur, however. In the event of a fire, the firefighter who has noticed and remembered that chemicals stored near the site of the fire might give off toxic fumes when burned will be the better prepared and more effective firefighter. The firefighter who has noticed and remembered the locations of fire stairs and firehose connectors will be on the scene more quickly and will attack the fire with the best efficiency. Should a firefighter find him- or herself inside a premises engulfed in dense smoke, memory of layout and points of exit can save his or her life.

Because observation and memory are so important to firefighters, many firefighter exams try to measure this ability in their applicants. The usual method of measuring this skill is to distribute a diagram or a picture (or perhaps two diagrams or pictures) to applicants to study for a given period of time. After the study period, usually five minutes per picture or diagram, the pictorial materials are collected. The applicant then must answer a series of multiple-choice questions based on information in the pictures or diagrams.

The pictures and diagrams on firefighter exams usually are not too complex. Five minutes is adequate for absorbing the information provided that the five minutes are used systematically. Your problem is that you cannot write anything down. All notes must be mental notes. It is worthwhile, therefore, to establish in advance the categories into which you will fit information and then to approach the pictorial material to fill in the blanks of each category.

1. *People.* Are there any people in the picture or diagram? If not, proceed to the next category. If so, where are they located? What landmark objects (windows, doors, fire escapes, fire sources) are they near? Are they in danger?
2. *Fire activity.* Is there fire? Is there smoke? Where is the activity located? Again, where is the fire activity located with respect to landmarks? How does it endanger life and property? Is there a predictable direction in which it might spread? (In connection with this last point, note whether there is a directional indicator—N—S—E—W—anywhere on the page.)
3. *Layout.* If the picture is an outdoor scene, look for fire equipment such as alarm boxes and hydrants. Note what these items are nearest to and farthest from. Study heights of buildings, natures of buildings, and fire escapes. If you have in hand an indoor scene or floor plan, focus on adjacent rooms, means of entering one room from the next, and locations of doors, windows, and fire escapes. Count and try to remember numbers of doors and doorways and numbers of windows.

4. *Special details.* Is there a smoke alarm? Where? Is there an obstruction noted? Is there an indication that doors are open or closed? If so, which ones? What is unique to this particular scene or diagram? What feature would likely lead to a question on an exam?

Give yourself five minutes to study each of the following drawings and diagrams, doing one exercise at a time. Study one diagram for five minutes, set it aside, answer the multiple-choice questions, and then study the answer discussion before going on to the next exercise.

Exercise One—Diagram

Exercise One—Questions

1. The main entry to these premises is the
   (A) door to the kitchen
   (B) front vestibule
   (C) door to the living room
   (D) door to the dining area

2. Direct access to the bathroom can be made from
   (A) bedroom 1
   (B) the living room
   (C) bedroom 2
   (D) none of these

3. The room with the fewest windows is
   (A) bedroom 2
   (B) the bathroom
   (C) the dining area
   (D) the kitchen

4. The exposure with the greatest possible number of exits is the
   (A) north
   (B) south
   (C) east
   (D) west

5. Entry to the kitchen is possible from
   (A) the dining area only
   (B) the dining area and the hall
   (C) the dining area and the garage
   (D) the dining area, hall, and garage

6. In the case of an electrical fire, power should be shut off in
   (A) the garage
   (B) the kitchen
   (C) the hall
   (D) there is no way to tell

7. The stairway goes upstairs by way of
   (A) the garage
   (B) the hall
   (C) bedroom 2
   (D) there is no stairway

8. If there were an out-of-control fire spreading from the fireplace, a person's best escape from the bathroom would be
   (A) out the bathroom door to the hall, into bedroom 1, and out the south window
   (B) out the bathroom door to the hall, through the living room and dining area, through the kitchen, and out by way of the garage
   (C) out the bathroom door, down the hall, and out the front door
   (D) across the hall into bedroom 2 and out the window

9. Connecting rooms are
   (A) bedroom 1 and bedroom 2
   (B) bedroom 2 and bedroom 3
   (C) bedroom 3 and the dining area
   (D) the dining area and the kitchen

10. The premises diagrammed here most probably are
    (A) a single-family detached house
    (B) a semidetached house
    (C) an apartment in a low-rise building
    (D) an apartment in a high-rise building

Answer Key

1. C
2. D
3. B
4. A
5. C
6. A
7. D
8. C
9. D
10. A

Explanatory Answers

1. **(C)** The front door enters directly into the living room. The door to the kitchen from the garage is a secondary entrance. There is no anteroom indicated in front of the living room.

2. **(D)** The bathroom door opens only into the hallway.

3. **(B)** The bathroom has no windows at all. Each of the other areas has one window.

4. **(A)** Most directional markers indicate North as up. This marker is unusual in that North is pointing down. Reverse the diagram to make North point up and get your bearings. The front of the house is the northern exposure. There are three windows and a door along the front (north) of the house.

5. **(C)** The hall does not go through to the kitchen; however, entry to the kitchen can be made through either the dining area or the garage.

6. **(A)** The electrical panel is in the garage.

7. **(D)** There is no stairway.

8. **(C)** This question requires both memory of the layout and common sense. Exit is always faster and safer by way of a door than by way of a window if it is possible to reach a door. The fire is at the end of the living room farthest from the door. The short run out the door makes most sense.

9. **(D)** Of these, only the dining area and kitchen connect.

10. **(A)** This takes some inference in conjunction with memory. The presence of a garage entry directly to the kitchen rules out an apartment house of any height. Windows on all four sides indicate that the house is free-standing and fully detached. There are no windows in a party wall.

In studying this floor plan, you should have noticed:

1. *People*. None.
2. *Fire activity*. None.
3. *Layout*. The northern orientation of house. Front entry to the living room. L-shaped living/dining area. Access to the bathroom from the hall only. Three nonconnecting bedrooms in a row. Corner bedrooms with two windows each. The hallway does not go through to the kitchen. The door from the dining area to the kitchen (as opposed to a possible open doorway without a door). Access to the kitchen through the garage. Fully interior bathroom.
4. *Special details*. Electrical panel in the garage. Fireplace in the living room. No fire-protective devices.

Exercise Two—Diagram

Exercise Two—Questions

1. To directly attack the fire, firefighters should
   (A) attach hoses to the hydrant next to the front door, enter through the living room, and go down the hall
   (B) attach hoses to the hydrant outside the kitchen door, running them out the kitchen window and into the bedroom window
   (C) attach hoses to the hydrant outside the bedroom and run them directly into the bedroom window
   (D) hook up to the standpipe in the hall and enter the bedroom

2. Closed doors retard the spread of fire and smoke. The door that should be closed right away is the
   (A) bathroom door
   (B) door to bedroom 2
   (C) door to bedroom 3
   (D) kitchen door

3. The cause of the fire most likely is
   (A) smoking in bed
   (B) electrical
   (C) a kitchen flare-up
   (D) unknown

4. There are two ways out of every room except
   (A) bedroom 2
   (B) the dining area
   (C) the bathroom
   (D) bedroom 3

5. Smoke detectors are in place in
   (A) the hall
   (B) the hall and the kitchen
   (C) the hall and the dining area
   (D) nowhere in this unit

6. In a search for people who might need rescue, firefighters should especially concentrate in
   (A) the kitchen
   (B) bedroom 3
   (C) the bathroom
   (D) the dining area

7. Doors to the outside of this unit can be found on the
   (A) south and east walls
   (B) north and east walls
   (C) south and west walls
   (D) north, south, and east walls

8. The total number of people in the unit is
   (A) 0
   (B) 1
   (C) 2
   (D) 3

9. A room with one window is
   (A) bedroom 1
   (B) bedroom 3
   (C) the living room
   (D) the bathroom

10. If the fire were to extend throughout the unit, the risk of involving furnishings would be greatest in
    (A) bedroom 3
    (B) the living room
    (C) the dining area
    (D) the kitchen

Answer Key

1. C
2. B
3. A
4. C
5. A
6. B
7. A
8. D
9. B
10. B

Explanatory Answers

1. **(C)** For the most direct attack on the fire, attach to the hydrant directly outside the involved room. The hydrant outside the front door should be utilized immediately thereafter for protection of the other rooms in the unit. With careful observation and memory, you should be aware that there is no hydrant outside the kitchen door and that the device in the hall is a smoke detector, not a standpipe.

2. **(B)** The bathroom door and the door to bedroom 3 already are closed. Danger to the immediately adjacent bedroom 2 is far greater than danger to the remote kitchen.

3. **(A)** The point of origin of the fire is a bed, and the victim is in the bedroom. Suspicion of smoking in bed as the cause of the fire is legitimate.

4. **(C)** Only the bathroom has just one exit. Bedrooms 2 and 3 each have a door and a window, and the dining area is wide open to the living room and the kitchen and has window access as well.

5. **(A)** There is one smoke detector at the end of the hall.

6. **(B)** The victim on the floor in bedroom 1 and the person in the kitchen are quite obvious. Note the person hiding behind a chair in the corner of bedroom 3. Of course, the bathroom and dining area must be checked as well, but they offer fewer hiding places and require less concentrated attention.

7. **(A)** Look at the directional marker. The north wall of the unit is at the top of the page, so the doors are on the south and east walls.

8. **(D)** The three people are the victim in bedroom 1, the person standing in the kitchen, and the person hiding in bedroom 3.

9. **(B)** Even though it is a corner bedroom, bedroom 3 has only one window. Bedroom 1 and the living room both have two windows, and the bathroom has none.

10. **(B)** The living room is full of upholstered furniture. The other rooms are more sparsely furnished.

In studying this diagram, you should have noticed:

1. *People*. Three—one each in bedroom 1, bedroom 3, and the kitchen.
2. *Fire activity*. Point of origin in the bedroom.
3. *Layout*. Exit doors from the living room and kitchen. Three nonconnecting bedrooms in a row. Fully interior bathroom with exit to the hallway only. L-shaped living/dining area. The hallway does not go through to the kitchen. Locations of windows. General volume and placement of furniture. Open and closed doors.
4. *Special details*. Hydrants at the front door and outside the window of bedroom 1. Smoke detector in the hall. Position of the victim.

Exercise Three—Sketch

Exercise Three—Questions

1. The fire is in
   - (A) the building between the apartments and the boutique
   - (B) the building next to the drug store
   - (C) a bank
   - (D) the building between the pizza parlor and the apartments

2. Of the people who can be seen, those in the most immediate danger are
   - (A) the people on the roof of the hotel
   - (B) the residents of the apartment house
   - (C) the people who live above the drug store
   - (D) the person in the doorway of the boutique

3. If the hour were 8 AM, the *least* peril to persons would occur at the
   - (A) hotel
   - (B) building between the hotel and the drug store
   - (C) sidewalk
   - (D) building housing the drug store

4. The best spot from which the firefighters should train their hoses on the fire is the
   - (A) hydrant in front of the theater
   - (B) hydrant in front of the pizza parlor
   - (C) roof of the boutique
   - (D) hydrant in front of the apartment house

5. The building *least* endangered by this fire is
   - (A) the apartment house
   - (B) a bank
   - (C) the theater
   - (D) the pizza parlor

6. The people on the roof of the hotel should be told to
   - (A) stay there
   - (B) go inside and stay inside
   - (C) take the elevator down to the lobby and leave the hotel
   - (D) take the stairs down to the lobby and leave the hotel

7. The total number of people visible in the drawing is
   - (A) 7
   - (B) 9
   - (C) 11
   - (D) 12

8. If firefighters want to get to the roof of the burning building after the flames and smoke have been brought under control, their best route is
   (A) directly from the roof of the apartment house
   (B) from the roof of the hotel to the roof of the drug store to the roof of the burning building
   (C) from the roof of the building housing the drug store to the roof of the building next door to the roof of the burning building
   (D) from the roof of the bank to the roof of the theater to the roof of the burning building

9. The fire probably was reported from the fire alarm box
   (A) in front of the hotel
   (B) on the wall of the theater
   (C) in front of the pizza parlor
   (D) on the fire escape of the apartment house

10. The wind is blowing from the
    (A) west
    (B) east
    (C) north
    (D) south

Answer Key

1. A
2. C
3. B
4. D
5. B
6. D
7. C
8. C
9. A
10. B

Explanatory Answers

1. **(A)** The fire is in the pizza parlor, which is between the apartments and the boutique.

2. **(C)** The people who live in the apartments over the drug store are separated from the fire by only one building. With the wind blowing in their direction, they clearly are in greater danger than the people in the apartment house next to the fire but upwind of it. The person at street level can just walk away; the people on the hotel roof are quite a bit removed.

3. **(B)** The building between the hotel and the drug store is a theater. At 8 AM a theater is likely to be deserted, creating peril to property but not to life.

4. **(D)** From the hydrant in front of the apartment house, the firefighters can move in quite close behind the fire. This is an ideal location. They can approach the fire directly without having to force their way through heat, flame, and smoke. The hydrant in front of the theater is farther away, and the approach is from a less favorable direction. There is no hydrant in front of the pizza parlor itself. The roof of the boutique is too dangerous a spot from which to fight the fire at this active stage.

5. **(B)** The two banks are the two least endangered buildings. The bank beside the apartment house is separated from the fire by the apartment house, and it is away from the path of the fire. The bank next to the hotel is the building farthest from the fire and is separated from it by a couple of large, sturdy buildings.

6. **(D)** This judgment question must be based on your recall of the whole scene. The people on the hotel roof (and the people inside the hotel) are in no immediate danger, but smoke and fire can spread and precautions must be taken. With the emergency in the area, electrical interruption could occur at any time. People should avoid the elevator and evacuate the hotel by way of the stairs. The roof of the hotel offers a wonderful vantage point for watching the fire, but a fire is not a show, and the smoke eventually will reach them. They must go.

7. **(C)** Five people on the hotel roof + two pedestrians + two in the apartments over the drug store + one in the doorway of the boutique + one on the apartment house fire escape = 11.

8. **(C)** Consider relative heights of adjacent buildings in answering this question. The drop from the apartment house is too steep, especially onto a roof weakened by fire. The route from the drug store allows for a gradual, cautious approach.

9. **(A)** Did you notice the legend on the page? The fire alarm box in front of the hotel is the only one in the picture. The device on the theater is a fire horn.

10. **(B)** The flame and smoke, and therefore the wind, are blowing from the east towards the west.

In studying this sketch, you should have noticed:

1. *People.* Numbers, locations, and locations with reference to the fire.
2. *Fire activity.* Location and location with reference to landmarks. Direction. Extent of involvement.
3. *Layout.* Types of buildings. Sizes of buildings. Arrangement of buildings. Relationships between buildings in terms of size and types of occupancy.
4. *Special details.* Legend on page. Number and location of hydrants, fire alarm boxes, and fire horns.

Exercise Four—Diagram

Exercise Four—Questions

1. The room most poorly protected in case of fire is
   (A) 1
   (B) 4
   (C) 6
   (D) 8

2. The room with the most comprehensive fire protection is
   (A) 2
   (B) 5
   (C) 7
   (D) 8

3. Fire hoses can be found in
   (A) the stairways
   (B) the center hall
   (C) the center hall and room 6
   (D) rooms 7 and 8

4. A room without access to a fire escape is
   (A) 2
   (B) 3
   (C) 5
   (D) 8

5. Activities known to be fire hazards are carried on in
   (A) rooms 1 and 2
   (B) rooms 5 and 7
   (C) rooms 2 and 8
   (D) rooms 5 and 8

6. In the case of a fire in room 1, the best exit would be the
   (A) fire escape from room 3
   (B) elevators
   (C) north stairs
   (D) south stairs

7. The room in which a dangerous situation exists right now is
   (A) 2
   (B) 4
   (C) 6
   (D) 8

8. In the case of a fire, an alert would be sounded from
   (A) the east fire escape
   (B) room 3
   (C) the hall
   (D) the elevators

9. The intricate pattern of partitions would make quick exit most difficult from room
   (A) 1
   (B) 2
   (C) 6
   (D) 7

10. The total number of windows on this floor is
    (A) 14
    (B) 16
    (C) 18
    (D) 20

Answer Key

1. A
2. C
3. B
4. A
5. D
6. D
7. B
8. C
9. A
10. C

Explanatory Answers

1. **(A)** Room 1 has no fire-protective device of any kind and has no independent exit to the outside. Room 6 is not much better off, but at least it has a heat detector. Room 4 has a fire escape but no internal protective device. Room 8 is well-protected.

2. **(C)** Room 7 has automatic sprinklers, a heat and smoke detector, a fire extinguisher, and a fire escape. Room 2 has only automatic sprinklers, and room 5 has only a heat and smoke detector. Room 8 has a smoke detector, a fire extinguisher, and a fire escape but lacks a sprinkler system.

3. **(B)** Fire hoses are only in the hallway.

4. **(A)** Room 2 has no access to a fire escape. Rooms 3 and 8 have fire escapes, and room 5 has a door connecting it with room 7, which has a fire escape.

5. **(D)** Use and storage of cleaning solvents and woodworking are activities that can pose fire hazards. The usual activities in offices and accessory showrooms do not present fire hazards. The nature of the business in the other areas is unspecified.

6. **(D)** Room 1 is not connected to a fire escape, so its occupants would have to go into the hall to escape the fire. Once in the hall, the quickest way out would be via the nearest (the south) stairway. Entering room 3 to go out the window and down the fire escape would be slower and more cumbersome. In addition, room 3 is the next room to become involved in the fire. Elevators should be avoided in a fire situation.

7. **(B)** There is a wastebasket afire in room 4.

8. **(C)** There is a fire alarm box topped by a bell in the hall next to the south staircase. The device in room 3 is a fire extinguisher, not an alarm.

9. **(A)** Room 1 is full of half partitions.

10. **(C)** Be sure to count the windows connected to fire escapes. There are 18 windows in all.

In studying this diagram, you should have noticed:

1. *People*. None.
2. *Fire activity*. Wastebasket fire in room 4.
3. *Layout*. Center core building with four elevators and two sets of stairs. All rooms have doors to the hallway and at least one window. Corner rooms have multiple windows. Rooms 5 and 7 connect. Rooms 3, 4, 7, and 8 have fire escapes.
4. *Special details*. Two fire hoses, a fire extinguisher, and a fire alarm box with bell in the hall. Fire-hazardous activities in rooms 5 and 8. Fire extinguishers in rooms 3, 7, and 8 and in the hall. Sprinkler systems in rooms 2 and 7. Heat detector in room 6. Smoke detector in room 8. Heat and smoke detectors in rooms 5 and 7. Diagram is of the third floor of the building. Extensive legend.

Exercise Five—Sketch

Exercise Five—Questions

1. The number of vehicles involved in the accident is
   - (A) 2
   - (B) 3
   - (C) 4
   - (D) 5

2. The number of known victims is
   - (A) 4
   - (B) 5
   - (C) 6
   - (D) 7

3. An extra hazard is being created by the
   - (A) pickup truck
   - (B) station wagon
   - (C) pedestrian in front of house 10
   - (D) person pulling the fire alarm

4. When firefighters arrive at the scene, they will be hampered by the
   - (A) situation at the west end of the street
   - (B) scarcity of fire hydrants
   - (C) absence of witnesses
   - (D) crowd of spectators

5. The house that is missing its house number is
   - (A) at the east end of the block
   - (B) in the middle of the block
   - (C) at the west end of the block
   - (D) house 8

6. The accident was probably caused by
   - (A) the station wagon
   - (B) the pickup truck
   - (C) the car
   - (D) a pedestrian

7. The animals on the scene are
   - (A) two dogs and a cat
   - (B) two cats and a dog
   - (C) three cats and a dog
   - (D) two cats and two dogs

8. The house that is different from the others is house
   - (A) 4
   - (B) 6
   - (C) 10
   - (D) all are exactly alike

9. The vehicle that is totally overturned is the
   - (A) station wagon
   - (B) sedan
   - (C) pickup truck
   - (D) convertible

10. When firefighters arrive on the scene, they must first direct their attention to the
    - (A) victims
    - (B) houses in danger of catching fire
    - (C) cars on fire
    - (D) bystanders

Answer Key

1. B
2. C
3. B
4. A
5. C
6. A
7. B
8. D
9. C
10. A

Explanatory Answers

1. **(B)** Three vehicles are piled up in the middle of the street. The other two cars are parked.

2. **(C)** Six people are lying in the street. We have no way of knowing how many are in the cars.

3. **(B)** The puddle forming from the back end of the station wagon most likely is gasoline. With the three vehicles actively on fire, a gasoline puddle creates a serious explosion threat.

4. **(A)** At the west end of the street, two cars are parked directly in front of and completely blocking access to the fire hydrant. This is both dangerous and illegal.

5. **(C)** House 2 at the far west end of the block is missing its house number.

6. **(A)** Note the one-way signs at each end of the street. It is possible that one or more of the vehicles spun around on impact, but on the face of the situation, it appears that the station wagon caused the accident by driving the wrong way on a one-way street.

7. **(B)** There are two cats and a dog in the scene. The animals have no bearing on the fire problem, but an observant test-taker must notice them.

8. **(D)** All the houses are exactly alike.

9. **(C)** The pickup truck has completely overturned and is resting on the top of its cab. The other two vehicles are more or less upright.

10. **(A)** Saving lives is more important than saving property. The bystanders must get out of the way by themselves; the victims need assistance.

In studying this sketch, you should have noticed:

1. *People.* Six lying in the street, one turning in a fire alarm at the street box, five approaching the scene.
2. *Fire activity.* Three vehicles apparently on fire after a collision.
3. *Layout.* Three cars in a pileup in the middle of street, one upside-down. Five identical houses in a row. Two parked cars blocking one hydrant. Second hydrant at the end of the street. Fire alarm box on corner.
4. *Special details.* Puddle coming from the rear end of the station wagon. An overturned pickup truck. One-way street, eastbound. Two cats and one dog.

MECHANICAL REASONING

The following questions will evaluate your understanding of some basic mechanical principles. Some questions will simply ask you to identify a pictured tool or to specify the tool's function. Other questions will ask you to predict the outcomes of mechanical activities or to reason backwards from an event to a probable mechanical cause.

   The firefighter exam will not assume that you have knowledge about specific firefighting equipment, tools, or techniques. It will assume, however, that you have a basic understanding of things that are mechanical in nature. Firefighters rely heavily on the machinery, tools, and safety equipment that help them perform their job successfully and safely. Therefore, it is important for a firefighter to be able to predict the outcome of a mechanical activity or to find the mechanical cause of an event.

   The following questions will provide you with a broad selection of the types of mechanical reasoning questions you might encounter on the firefighter exam. A full set of explanations will follow the questions.

Directions: *Choose the best answer and circle the corresponding letter.*

1. The main reason a fiberglass ladder would be used in place of an aluminum ladder is when there are concerns about
   (A) the length of the ladder
   (B) the conductivity of the ladder
   (C) the flammability of the ladder
   (D) the "rust-resistance" of the ladder

2. The main advantage of four-wheel drive is
   (A) higher speed
   (B) better traction
   (C) better gas mileage
   (D) greater durability of the vehicle

3. One complete revolution of the sprocket wheel will bring weight W2 higher than weight W1 by
   (A) 20 inches
   (B) 40 inches
   (C) 30 inches
   (D) 50 inches

4. If you are cooking on your stove at home and the grease in your frying pan catches fire, the method that is NOT proper for putting out the fire is
   (A) extinguishing the fire with carbon dioxide
   (B) covering the pan with its lid
   (C) pouring water on the fire
   (D) extinguishing the fire with a noncombustible foam

5. In the figure shown here, assume that all valves are closed. For air to flow from R through G and then through S to M, open

    (A) valves 1, 2, 6, and 4
    (B) valves 7, 3, and 4
    (C) valves 7, 6, and 4
    (D) valves 7, 3, and 5

6. Which of the following would be best for cutting wire?

    (A)

    (B)

    (C)

    (D)

7. The function of the generator or alternator is to
    (A) start the engine
    (B) carry electricity from the battery to the engine
    (C) keep the battery charged
    (D) control the production of hydrocarbons

8. The principal objection to using water from a hose to put out a fire involving electrical equipment is that
    (A) serious shock might result
    (B) it might spread the fire
    (C) metal parts might rust
    (D) fuses might blow out

9. The following figure shows four gears. If gear 1 turns as shown, the gears turning in the same direction are
   (A) 2, 3, and 4
   (B) 2 and 4
   (C) 2 and 3
   (D) 3 and 4

10. Light fixtures suspended from chains should be wired so that the
    (A) wires do not support the fixture
    (B) wires help support the fixture
    (C) chains have an insulated link
    (D) chain is not grounded to prevent short circuits

11. Hose X is 50 feet long and $\frac{1}{2}$ inch wide, and hose Y is 50 feet long and 1 inch wide. If they begin filling at the same time and an equal amount of water pressure is used to fill each hose,
    (A) hose X will fill with water first
    (B) hose Y will fill with water first
    (C) they both will fill with water at the same time
    (D) there is no way of knowing which hose will fill first

12. In the figure on the right, all four springs are identical. In case 1 with the springs end to end, the stretch of each spring caused by the five-pound weight is
    (A) one-half as much as in case 2
    (B) the same as in case 2
    (C) twice as much as in case 2
    (D) four times as much as in case 2

13. You might use this instrument if you wanted to
    (A) pitch a tent
    (B) poke holes in a fabric
    (C) locate studs in a wall
    (D) drill holes at equal short distances along a board

14. After using cotton rags to clean a spill with paint thinner, you should
    (A) leave them in a pile
    (B) stuff them all in a metal can
    (C) stuff them all in a plastic container
    (D) hang them separately to dry

15. One complete revolution of the following windlass drum will move the weight up
    (A) $\frac{1}{2}$ foot
    (B) $1\frac{1}{2}$ feet
    (C) 1 foot
    (D) 2 feet

16. The frequency of oiling and greasing bearings and other moving parts of machinery depends mainly on the
    (A) size of the parts requiring lubrication
    (B) speed at which the parts move
    (C) ability of the operator
    (D) amount of use of the equipment

17. The tool shown here is a
    (A) screw
    (B) screwdriver
    (C) drill bit
    (D) corkscrew

18. If pipes A and B are free to move back and forth but are held so they cannot turn and the coupling is turned four revolutions with a wrench, the overall length of the pipes and coupling will
    (A) decrease $1/2$ inch
    (B) remain the same
    (C) increase or decrease 1 inch depending on the direction of turning
    (D) increase $1/2$ inch

19. The plug of a portable tool should be removed from the convenience outlet by grasping the plug, not by pulling on the cord. This is because
    (A) the plug is easier to grip than the cord
    (B) pulling on the cord might cause the plug to fall on the floor and break
    (C) pulling on the cord might break the wires off the plug terminals
    (D) the plug generally is better insulated than the cord

20. The following figure is a device that is attached to the top of a ladder. Its purpose is
    (A) to allow one to hang the ladder for easier storage
    (B) to increase the strength of the ladder
    (C) to give the ladder a greater amount of stability
    (D) to allow the ladder to be used upside down

21. The figure represents an enclosed water chamber that is partially filled with water. The number 1 indicates air in the chamber, and 2 indicates a pipe by which water enters the chamber. If the water pressure in the pipe (2) increases, then the
    (A) water pressure in the chamber will be decreased
    (B) water level in the chamber will fall
    (C) air in the chamber will be compressed
    (D) air in the chamber will expand

22. Products stored in aerosol cans often have warnings against storing such containers in areas that might exceed 120°F. The main reason for these warnings is that
    (A) high temperatures might cause the contents of the container to shrink
    (B) high temperatures might cause the contents to shift in the container
    (C) high temperatures might cause the contents to separate
    (D) high temperatures might cause the contents of the container to expand

23. When removing the insulation from a wire before making a splice, care should be taken to avoid nicking the wire mainly because the
    (A) current carrying capacity will be reduced
    (B) resistance will be increased
    (C) wire tinning will be injured
    (D) wire is more likely to break

24. Wood screws properly used as compared to nails properly used
    (A) are easier to install
    (B) generally hold better
    (C) are easier to drive flush with the surface
    (D) are more likely to split the wood

25. If a hose releases water at 2,500 PSI, the pressure of the water coming out of the hose is
    (A) greater than if it were 1,500 psi
    (B) less than if it were 1,500 psi
    (C) about the same as if it were 1,500 psi
    (D) close to that found in a household sink

26. The figure shows a cutter and a steel block. For proper cutting, they should move respectively in directions
    (A) 1 and 4
    (B) 2 and 3
    (C) 1 and 3
    (D) 2 and 4

27. The reading shown on the gage is
    (A) 10.35
    (B) 13.5
    (C) 10.7
    (D) 17.0

28. An electric light bulb operated at *more* than its rated voltage will result in a
    (A) longer life and dimmer light
    (B) longer life and brighter light
    (C) shorter life and brighter light
    (D) shorter life and dimmer light

29. The wrench used principally for pipe work is
    (A)

    (B)

    (C)

    (D)

30. A slight coating of rust on small tools is best removed by
    (A) applying a heavy coat of vaseline
    (B) rubbing with kerosene and fine steel wool
    (C) scraping with a sharp knife
    (D) rubbing with a dry cloth

31. Which of the following is most likely to have difficulty stopping at high speeds?

(A)

(B)

(C)

(D)

32. To bring the level of the water in the tanks to a height of $2\frac{1}{2}$ feet, the quantity of water to be added is
  (A) 10 quarts
  (B) 20 quarts
  (C) 15 quarts
  (D) 25 quarts

33. The outlet that will accept the plug is
  (A) 1
  (B) 2
  (C) 3
  (D) 4

34. The micrometer shown here reads
    (A) .2270
    (B) .2120
    (C) .2252
    (D) .2020

35. When a spark plug ignites fuel in the combustion chamber,
    (A) you should stop the car immediately
    (B) the spark plug is functioning properly
    (C) the battery is not working
    (D) the vehicle is out of gas

36. Which of the following statements is true?
    (A) If the nut is held stationary and the head is turned clockwise, the bolt will move up.
    (B) If the head of the bolt is held stationary and the nut is turned clockwise, the nut will move down.
    (C) If the head of the bolt is held stationary and the nut is turned clockwise, the nut will move up.
    (D) If the nut is held stationary and the bolt is turned counterclockwise, the nut will move up.

37. A good lubricant for locks is
    (A) graphite
    (B) grease
    (C) mineral oil
    (D) motor oil

38. The best way to neutralize an acid is to
    (A) mix it with an acid of a weaker strength
    (B) mix it with a neutral substance
    (C) mix it with a base of equal strength
    (D) mix it with a stronger acid

39. The tool shown here is a
    (A) punch
    (B) drill holder
    (C) Phillips-type screwdriver
    (D) socket wrench

40. Wood ladders should not be painted because
    (A) the paint will wear off rapidly due to the conditions under which ladders are used
    (B) ladders are slippery when painted
    (C) it is more effective to store the ladder in a dry place
    (D) paint will hide defects in the ladder

41. If pipe A is held in a vise and pipe B is turned 10 revolutions with a wrench, the overall length of the pipes and coupling will decrease
    (A) $5/8$ inch
    (B) $2\frac{1}{2}$ inches
    (C) $1\frac{1}{4}$ inches
    (D) $3\frac{3}{4}$ inches

42. Which of the following saws is used to make curved cuts?
    (A)

    (B)

    (C)

    (D)

43. A method that can be used to prevent the formation of "skin" on a partially used can of oil paint is to
    (A) turn the can upside down every few months
    (B) pour a thin layer of solvent over the top of the paint
    (C) store the paint in a well-ventilated room
    (D) avoid shaking the can after it has been sealed

44. The reason that a lubricant prevents rubbing surfaces from becoming hot is that the oil
    (A) is cold and cools off the rubbing metal surfaces
    (B) is sticky, preventing the surfaces from moving over each other too rapidly
    (C) forms a smooth layer between the two surfaces, preventing them from coming into contact
    (D) makes the surfaces smooth so they move easily over each other

45. The tool shown here is used to
    (A) set nails
    (B) drill holes in concrete
    (C) cut a brick accurately
    (D) centerpunch for holes

46. If the block on which the lever is resting is moved closer to the brick,
    (A) the brick will be easier to lift and will be lifted higher
    (B) the brick will be harder to lift and will be lifted higher
    (C) the brick will be easier to lift but will not be lifted as high
    (D) the brick will be harder to lift and will not be lifted as high

47. When driving a nail into a piece of wood, the most effective way to prevent the wood from splitting is to
    (A) use a carpenter's hammer
    (B) first drill a small hole in the wood
    (C) drive the nail into the wood slowly
    (D) drive the nail into the wood quickly

48. If the head of a hammer has become loose on the handle, it should be properly tightened by
    (A) driving the handle further into the head
    (B) driving a nail alongside the present wedge
    (C) using a slightly larger wedge
    (D) soaking the handle in water

49. The following figure represents a water tank containing water. The number 1 indicates an intake pipe, and the number 2 indicates a discharge pipe. Of the following, the *least* accurate statement is that the
    (A) tank will eventually overflow if water flows through the intake pipe at a faster rate than it flows through the discharge pipe
    (B) tank will empty completely if the intake pipe is closed and the discharge pipe is allowed to remain open
    (C) water in the tank will remain at a constant level if the rate of intake is equal to the rate of discharge
    (D) water in the tank will rise if the intake pipe is operating when the discharge pipe is closed

50. The tool shown here is most often used for
   (A) cutting metal
   (B) cutting wood
   (C) cutting tile
   (D) cutting glass

51. The tool shown here is
   (A) an Allen-head wrench
   (B) a double scraper
   (C) an offset screwdriver
   (D) a nail puller

52. Wires often are spliced using a fitting like the one shown here. The use of this fitting does away with the need for
   (A) skinning
   (B) cleaning
   (C) twisting
   (D) soldering

53. With which of the following screw heads do you use an Allen wrench?

   (A)

   (B)

   (C)

   (D)

54. A trowel is a tool used for
   (A) clearing out drain pipes
   (B) digging holes for footings
   (C) carving notches into wood
   (D) smoothing out cement

55. The main reason some nails are galvanized is to
   (A) make them stronger
   (B) make the point sharper
   (C) make them smoother
   (D) make them rust resistant

56. When driving in extremely low temperatures, a car's engine might overheat if
   (A) the heater is on for too long
   (B) the coolant freezes
   (C) the gas tank is not full enough
   (D) the oil recently was changed

57. Which of the following would be best for digging into the ground?

   (A)

   (B)

   (C)

   (D)

58. The easiest way to chop through wood with an ax is to align the blade
    (A) with the grain
    (B) across the grain at 90°
    (C) across the grain at 45°
    (D) around the grain

59. The convenience outlet known as a *polarized* outlet is number
    (A) 1
    (B) 2
    (C) 3
    (D) 4

60. The tool used to measure the depth of a hole is

    (A)

    (B)

    (C)

    (D)

61. Neutral wire can quickly be recognized by the
    (A) greenish color
    (B) bluish color
    (C) natural or whitish color
    (D) black color

62. If a wrench were attached to the top of a fire hydrant, as pictured here, how would you turn the water on?
    (A) pull the wrench up
    (B) turn the wrench counterclockwise
    (C) push the wrench down
    (D) turn the wrench clockwise

63. If the flush tank of a water-closet fixture overflows, the fault is likely to be
- (A) failure of the ball to seat properly
- (B) excessive water pressure
- (C) a defective trap in the toilet bowl
- (D) a waterlogged float

64. When using a hammer with one hand, to use the greatest amount of force, you should hold the hammer
- (A) near the end of the handle
- (B) in the middle of the handle
- (C) near the head of the hammer
- (D) by the head of the hammer

65. Study the gear wheels in the figure shown here and then determine which of the following statements is true.
- (A) If you turn wheel M clockwise by means of the handle, wheel P also will turn clockwise.
- (B) It will take the same time for a tooth of wheel P to make a full turn as it will for a tooth of wheel M.
- (C) It will take less time for a tooth of wheel P to make a full turn than it will take a tooth of wheel M.
- (D) It will take more time for a tooth of wheel P to make a full turn than it will for a tooth of wheel M.

66. Locknuts frequently are used in making electrical connections on terminal boards. The purpose of the locknuts is to
- (A) eliminate the use of flat washers
- (B) prevent unauthorized personnel from tampering with the connections
- (C) keep the connections from loosening through vibration
- (D) increase the contact area at the connection point

67. When dragging a slightly leaky, cloth-covered hose filled with water across a gravel surface, the element most likely to make the task difficult is
    (A) the cloth hose covering
    (B) the gravel surface
    (C) the leaking water
    (D) the weight of the water

68. The tool shown here is used to
    (A) ream holes in wood
    (B) countersink holes in soft metals
    (C) turn Phillips-head screws
    (D) drill holes in concrete

69. The tool shown here is used for
    (A) soldering
    (B) caulking
    (C) shooting
    (D) scoring

70. A squeegee is a tool used in
    (A) drying windows after washing
    (B) cleaning inside boiler surfaces
    (C) the central vacuum cleaning system
    (D) clearing stoppage in waste lines

71. Boxes and fittings intended for outdoor use should be
    (A) of weatherproof type
    (B) stamped steel of not less than No. 16
    (C) standard gauge
    (D) stamped steel plated with cadmium

72. The device used to change AC to DC is a
    (A) frequency changer
    (B) regulator
    (C) transformer
    (D) rectifier

73. The figure shown here shows a governor on a rotating shaft. As the shaft speeds up, the governor balls will move
    (A) down
    (B) upward and inward
    (C) upward
    (D) inward

74. The purpose of an air valve in a heating system is to
    (A) prevent pressure from building up in a room due to the heated air
    (B) relieve the air from steam radiators
    (C) allow excessive steam pressure in the boiler to escape to the atmosphere
    (D) control the temperature in the room

75. If a fuse of higher than the required current rating is used in an electrical circuit,
    (A) better protection will be afforded
    (B) the fuse will blow more often because it carries more current
    (C) serious damage might result to the circuit from overload
    (D) maintenance of the large fuse will be higher

76. If the following dark lines represent different ways a ladder can be set up, which line represents the safest way to set up a ladder?

    (A)

    (B)

    (C)

    (D)

77. Paint is "thinned" with
    (A) linseed oil
    (B) varnish
    (C) turpentine
    (D) gasoline

78. The tool shown here is
    (A) an offset wrench
    (B) a box wrench
    (C) a spanner wrench
    (D) an open end wrench

79. If gear A makes one clockwise revolution per minute, which one of the following is true?
    (A) Gear B makes one counterclockwise revolution every 4 minutes.
    (B) Gear C makes two clockwise revolutions every minute.
    (C) Gear B makes four clockwise revolutions every minute.
    (D) Gear C makes one counterclockwise revolution every 8 minutes.

80. Referring to the figure, which one of the following statements is true?
    (A) If the nut is held stationary and the head is turned clockwise, the bolt will move down.
    (B) If the head of the bolt is held stationary and the nut is turned clockwise, the nut will move down.
    (C) If the head of the bolt is held stationary and the nut is turned clockwise, the nut will move up.
    (D) If the nut is held stationary and the head is turned counterclockwise, the bolt will move up.

81. If the float of a flush tank leaks and fills with water, the most probable result is that
    (A) there will be no water in the tank
    (B) the ball cock will remain open
    (C) water will flow over tank rim onto floor
    (D) the flush ball will not seat properly

82. If the temperature gauge indicates that the engine is getting overheated,
    (A) allow it to cool down
    (B) pour cold water in immediately
    (C) pour hot water in immediately
    (D) pour in a cooling antifreeze at once

83. A crack in the distributor cap might cause problems by
    (A) allowing oil to leak out
    (B) allowing electricity to leak out
    (C) allowing fuel to seep in
    (D) allowing moisture to seep in

84. The sketch here shows a head-on view of a three-pronged plug used with portable electrical power tools. Considering the danger of shock when using such tools, it is evident that the function of the U-shaped prong is to
    (A) ensure that the other two prongs enter the outlet with the proper polarity
    (B) provide a half-voltage connection when doing light work
    (C) prevent accidental pulling of the plug from the outlet
    (D) connect the metallic shell of the tool motor to ground

85. The object pictured here is a
    (A) spark plug
    (B) lug nut
    (C) drill bit
    (D) crank shaft

Answer Key

1. B	30. B	59. A
2. B	31. D	60. C
3. B	32. C	61. C
4. C	33. C	62. B
5. D	34. A	63. D
6. A	35. B	64. A
7. C	36. B	65. D
8. A	37. A	66. C
9. D	38. C	67. D
10. A	39. D	68. C
11. A	40. D	69. B
12. C	41. C	70. A
13. D	42. B	71. A
14. D	43. B	72. D
15. C	44. C	73. C
16. D	45. B	74. B
17. C	46. C	75. C
18. B	47. B	76. C
19. C	48. C	77. C
20. C	49. B	78. D
21. C	50. A	79. B
22. D	51. C	80. C
23. D	52. D	81. B
24. B	53. C	82. A
25. A	54. D	83. D
26. C	55. D	84. D
27. D	56. B	85. A
28. C	57. C	
29. C	58. C	

Explanatory Answers

1. **(B)** The main reason a fiberglass ladder would be used in place of an aluminum ladder is if there were concerns about conductivity or the risk of electrical shock from nearby electrical wires. A fiberglass ladder does not have any advantages in length (A) over an aluminum ladder. Fiberglass and aluminum are not flammable (C) nor are they susceptible to rusting (D).

2. **(B)** Because power from the engine is transmitted to both the front wheels and the back wheels, a four-wheel drive vehicle has much better traction. Four-wheel drive is especially useful in mud, sand, and very uneven terrain. Although four-wheel drive tends to be a feature of durable vehicles (D) such as jeeps, it is not the four-wheel drive that makes them durable.

3. **(B)** One-half the circumference of the sprocket wheel is 10 inches; therefore, the entire circumference is 20 inches. In one complete revolution of the wheel, the chain will move 20 inches. As weight 2 moves up 20 inches, weight 1 will move down 20 inches. The difference between the heights of the two weights will be 40 inches.

4. **(C)** Water and grease do not mix. Therefore, if you pour water on a grease fire, it might cause the grease to splatter and actually spread the fire. Fire feeds off oxygen, so extinguishing it with carbon dioxide would work (A). Covering the pan would cut off the oxygen supply (B), and a noncombustible foam also would cut off the oxygen supply (D).

5. **(D)** The air from R must follow a route down through valve 7, up through G, through valve 3, down through S, to the right through 5, and then up and over to M. The air could not pass through valves 1 and 2 to G because it would escape through the opening between valves 1 and 2. If either valve 4 or valve 6 were to be opened, the air would be diverted from the appointed route.

6. **(A)** This is a wire cutter. (B) is a pruning clipper. (C) is a hedge clipper. (D) is a pair of pliers.

7. **(C)** The generator or alternator is operated by the car's engine. It produces electricity that flows to the battery and keeps the battery charged.

8. **(A)** Water is not likely to spread an electrical fire (B), though it might be ineffective at putting it out. It is true that metal parts might rust if soaked with water (C) and that fuses might blow (D), but these are not serious considerations. There is a real danger from fatal shock if water is used on an electrical fire. The danger is even greater if that water comes from a hose because there would then be a great deal of water on the floor and the firefighters would be standing in it. Electricity travels rapidly through water, and the people standing in the water would be its targets.

9. **(D)** A turning gear always turns the gear with which it interlocks in the opposite direction. If gear 1 turns clockwise, gear 2 must turn counterclockwise. In turn, gears 3 and 4, because they are both turned by gear 2, must both turn clockwise.

10. **(A)** The answer to this question is pure common sense. Electrical wires should serve only one purpose—to supply electricity. If electrical wires are required to support weight, there is danger of breakage in the wires and of damage to the fixture itself caused by tension at the connections.

11. **(A)** Hose X and hose Y are the same length and are filled starting at the same time with the same amount of water pressure. Hose X, however, is only $\frac{1}{2}$ inch wide; hose Y is 1 inch wide. This means that hose X has less space inside it and will fill up faster than hose Y.

12. **(C)** In case 2, each spring bears one-half the weight of the 5-pound weight. Each spring in case 2 is therefore stretched by $2\frac{1}{2}$ pounds. In case 1, each spring bears a full 5-pound load. Each spring in case 1 must be stretched twice as much as each spring in case 2.

13. **(D)** The compass shown would be very useful in marking out equal short distances on a board. The compass would not, of course, be of use in the actual drilling.

14. **(D)** Rags used to clean a spill with paint thinner should always be allowed to dry after use. Leaving them in a pile (A), stuffing them in a metal can (B), or stuffing them in a plastic container (C) might cause the rags to catch fire due to the paint thinner on them.

15. **(C)** Because the circumference of the drum is one foot, one complete revolution of the drum will take up one foot of each rope. As each of the separate ropes supporting the weight is shortened by one foot, the weight will move up one foot.

16. **(D)** Lubrication of machinery is scheduled according to time elapsed and amount of use. All the other reasons offered are irrelevant.

17. **(C)** The tool is a drill bit. The smooth end is inserted into the drill chuck and is tightened so it is very secure. When the drill motor goes, the bit turns very fast and drills holes wherever it is applied.

18. **(B)** If the coupling is turned but the pipes are held firm so they cannot turn, the coupling will move along the length of one or the other of the pipes, but the overall length of the three pieces will remain the same.

19. **(C)** Yanking at an electric cord can cause hidden damage inside the plug. The force can cause terminals to loosen or bits of wire to break off and fray. Frayed wires can, in turn, come into contact with the opposite pole, causing a fire or short circuit in the plug at some later date.

20. **(C)** The figure is a picture of a stabilizer. This device is attached to the top of a ladder to give it a greater amount of stability by reducing the likelihood that the ladder will slide to the left or the right.

21. **(C)** If water pressure in the pipe is increased, more water will flow into the water chamber. Because the chamber is enclosed, the air will be unable to escape. As more water enters the chamber, the existing air must be compressed into a smaller space.

22. **(D)** The main reason for storing aerosol cans in temperatures that do not exceed 120°F is because high temperatures might cause the contents of the can to expand. This might then cause the can to burst.

23. **(D)** A nick in a wire can be dangerous because it weakens the wire at that point and can lead to breakage. If the wire is nicked during stripping, you should cut off the weakened portion and begin again. Later breakage from an unnoticed weakness can lead to a short circuit.

24. **(B)** Wood screws are usually more difficult to install than nails, but they often are preferable because they generally hold better and are less likely to split the wood.

25. **(A)** PSI refers to how much water pressure is present in pounds per square inch. Therefore, 2,500 PSI is much greater than 1,500 psi. This amount of water pressure is far greater than that coming out of a household sink (D).

26. **(C)** Common sense should give you the answer to this question. For the cutter to cut the block, the two must be in contact. Obviously, the block must move in direction 3 to make contact with the cutter. For a cutter to cut, it must move in a direction that enables the sharp edge of its teeth to bite into the object being cut. In direction 1, the teeth will bite into the block. In direction 2, the back edge of the teeth will just slide off the block.

27. **(D)** Each division marks 2 units: 20 units/10 divisions = 2 units/division. The pointer is $3\frac{1}{2}$ divisions above 10: 2 units/division × $3\frac{1}{2}$ divisions = 7 + 10 = 17.

28. **(C)** If an electric light bulb is operated at more than its rated voltage, the extra surge of electricity will cause the bulb to burn more brightly. The same excess electrical force, however, will weaken the filament and will cause the bulb to burn out more quickly.

29. **(C)** This is a pipe wrench. (A) is a crescent or expandable wrench. (B) is a ratchet wrench. (D) is an open end wrench.

30. **(B)** Kerosene and fine steel wool will effectively remove a thin layer of rust. Vaseline (A) can prevent rust but not remove it. A sharp knife (C) might chip off thick crusty rust but not a thin coat. A dry cloth (D) alone will not remove rust.

31. **(D)** A larger vehicle is likely to be heavier and therefore will have more difficulty stopping at high speeds.

32. **(C)** Ten quarts of water have brought the water level to 1 foot. An additional 15 quarts would raise the water level by $1\frac{1}{2}$ feet to a total height of $2\frac{1}{2}$ feet.

33. **(C)** This shape plug will only go into a similarly shaped socket. The other outlets are of different shapes.

34. **(A)** The measurements that can be made on the micrometer are: 1) 2 major divisions and 1 minor division on the ruler-type scale: .2 + .025 = .225; 2) 2 minor divisions above 0 on the rotating scale: .002. Summing, we find the final measurement is .225 + .002 = .227.

35. **(B)** The main purpose of a spark plug is to ignite fuel vapors in the combustion chamber. The battery must be working for this to occur (C), the vehicle must have gas in it for this to occur (D), and this is definitely not a reason to stop the car (A).

36. **(B)** Clockwise is left to right. If the nut moves, it follows the threads of the bolt downward.

37. **(A)** Graphite, which is powdered carbon, is very slippery and will not bind the small springs and metal parts of a lock.

38. **(C)** A base is a substance that is the "opposite" of an acid. The best way to neutralize an acid is to mix it with a base of equal strength. An acid of weaker strength will not have as good of an effect (A). A neutral substance will work better than an acid, but it will not completely neutralize the acid (B). A stronger acid will only make the acid stronger.

39. **(D)** Although this tool looks like a screwdriver, the head will fit into a hex nut and works like a socket wrench.

40. **(D)** You cannot see a defect in a painted ladder such as a knot or a split in the wood. A ladder should *never* be painted.

41. **(C)** The overall length of the pipes and coupling could decrease or increase, depending on the direction in which pipe B is turned. As stated in this question, however, pipe B is turned to disappear into the coupling. Because there are eight threads to the inch, eight complete revolutions of the pipe would shorten the pipes and coupling by one inch. An additional two turns, for a total of ten, would shorten the pipes and coupling by an additional $\frac{2}{8}$ or $\frac{1}{4}$ of an inch.

42. **(B)** This is a keyhole saw used to make curved cuts. (A) is a backsaw. (C) is a rip or crosscut saw. (D) is a hacksaw.

43. **(B)** "Skin" forms when air combines with paint. To stop this from happening, pour a thin layer of solvent over the paint. The air will be prevented from reaching the paint and forming a layer of skin.

44. **(C)** When two pieces of metal rub together, the friction causes a great deal of heat. Oil reduces the friction between the two pieces of metal.

45. **(B)** The tool shown is a star drill. It is hit with a hammer to make a hole in concrete.

46. **(C)** If the block is moved toward the brick, the moment for a given force exerted will increase (being further from the force), making it easier to lift. The height will be made smaller, hardly raising the brick when moved to the limit (directly underneath it).

47. **(B)** By first drilling a small hole into the wood, you can reduce the amount of stress the nail places on the wood and can therefore reduce the likelihood that the wood will split.

48. **(C)** If you look at the top of a hammer where it is joined to the handle, you will see the top of either a wooden wedge or a metal wedge. Driving another wedge into the handle will tighten the hammer.

49. **(B)** If pipe 2 is open while pipe 1 is closed, the level will drop to the lowest level of pipe 2. This leaves the volume below pipe 2 still filled, having no way to discharge. All other statements are true.

50. **(A)** The tool shown is a hack saw, which is used to cut metal.

51. **(C)** This tool is an offset screwdriver. It is used for tightening screws in hard-to-reach places where a regular screwdriver cannot turn in a complete revolution.

52. **(D)** This is a mechanical or solderless connector. It does away with the need to solder wires and is found in house wiring.

53. **(C)** An Allen wrench is hexagon-shaped and will fit into screw C.

54. **(D)** A trowel is a tool with a flat face that often is used to smooth out cement.

55. **(D)** A galvanized nail is covered with a protective coating that keeps the nail from rusting.

56. **(B)** It seems strange to think that a car could overheat when it is cold outside; however, extremely cold temperatures can freeze the engine coolant. This makes the coolant useless and leads to an overheated engine.

57. **(C)** This type of shovel is best for digging into the ground. The other shovels are used for shoveling snow (A) and (D) and for shoveling grain (B).

58. **(C)** Wood is weakest when a cut is made with the grain. Any angle that is across the grain will not be as effective (B) and (C). There is no way to cut around the grain (D).

59. **(A)** The plug can go into the outlet in only one way in a polarized outlet. In the other outlets, the plug can be reversed.

60. **(C)** The flattened part rests at the top of the hole, and the ruler is pushed down into the hole until it reaches the bottom. The depth of the hole is then read from the ruler.

61. **(C)** Neutral wire is whitish in color. The hot lead is black, and the ground wire is green.

62. **(B)** Just as with any water valve, you should turn it counterclockwise to turn the water on. Pulling the wrench up (A) and pushing it down (C) will have no effect. Turning the wrench clockwise will actually tighten the valve.

63. **(D)** The water shut-off valve on a flush tank is closed by the force of a lightweight ball rising inside the tank. If this float becomes waterlogged, it will not rise and shut off the water.

64. **(A)** Holding the hammer near the end of the handle enables you to take advantage of the weight of the hammer with each swing and will result in a more forceful blow.

65. **(D)** Wheel P has 16 teeth; wheel M has 12 teeth. When wheel M makes a full turn, wheel P will still have four more teeth to turn. Therefore, wheel P is slower and takes more time to turn.

66. **(C)** Locknuts are bent so that their metal edges will bite into the terminal board and will require the use of a wrench to loosen them.

67. **(D)** The weight of the water is most likely to make the task difficult because it makes the hose extremely heavy and hard to drag. The cloth hose covering will have little effect (A), the gravel surface might actually make the task easier by providing a movable surface over which to drag the hose (B), and the leaking water will have little consequence (C).

68. **(C)** This is a Phillips-head screwdriver. It will turn Phillips-head screws with this shape:

69. **(B)** The tool pictured is a caulk gun and is used for caulking.

70. **(A)** A squeegee is a rubber wiper that removes water from a wet window.

71. **(A)** Outdoor boxes and fittings must be weatherproof to withstand any problems caused by moisture.

72. **(D)** A rectifier, or diode, is a device that changes AC to DC.

73. **(C)** The centrifugal force acts to pull the balls outward. Because the two balls are connected to a yolk around the center bar, this outward motion pulls the balls upward.

74. **(B)** An air valve on a radiator removes air from the steam pipes. If air is trapped in the pipes, it prevents the steam from going to the radiator. This prevents the radiator from producing heat.

75. **(C)** Never use a fuse with a rating higher than the one specifically called for in the circuit. A fuse is a safety device used to protect a circuit from serious damage caused by too high a current.

76. **(C)** This is the best angle for a ladder setup. Both (A) and (D) might cause a person standing at the top to fall over backwards. (B) might cause the bottom of the ladder to slip backwards, causing the ladder to fall.

77. **(C)** Paint is made thinner or easier to apply by diluting it with turpentine. Linseed oil and varnish are not used as paint thinners.

78. **(D)** The opened face on this tool shows that it is an open end wrench.

79. **(B)** Gear A turns in the opposite direction from gear B. A clockwise turn of A results in a counterclockwise revolution of B. Because the distance traversed by A (perimeter = × diameter = × 4) is twice that of C (perimeter = × 2), the speed of C is doubled.

80. **(C)** To tighten the bolt, turn it counterclockwise. To tighten the nut on the bolt, the reverse is true—turn it clockwise.

81. **(B)** The ball cock will remain open, the float will not rise with the incoming water, and the tank will continue to fill because the shut-off valve will not be shut off.

82. **(A)** If an engine starts to overheat, you must stop the car immediately. Otherwise, the metal will expand and damage the engine. If you pour anything on an overheated engine, the rapid cooling might cause the block to crack.

83. **(D)** A crack in the distributor cap might cause moisture to seep in and cause a short in the system. This might prevent the car from starting.

84. **(D)** The third prong in the plug is the grounding wire.

85. **(A)** The object pictured is a spark plug.

PART TWO

CONTENTS

Practice Examination 1   155
Answers and Explanations   200
Practice Examination 2   211
Answers and Explanations   233
Practice Examination 3   243
Answers and Explanations   265

PRACTICE EXAMINATION 1

The examination that follows is an actual firefighter examination given in New York City, and it is reprinted here by permission of the city. If you will be taking a New York City exam for firefighters, it will NOT be the same as this one. The format may be similar, but the questions will be different. If you plan to take a firefighter exam anywhere outside New York City, you cannot assume that your exam will look like this one at all. Practicing with this examination, however, still should provide excellent preparation for whatever your own town or city might have in store.

All of the instructions and cautions are included here along with the exam itself so you can see how seriously this exam must be taken. Keep in mind how important this exam is as you study and prepare yourself.

City of New York
Department of Personnel

EXAMINATION NO. 7022
FIREFIGHTER

Social Security No. _____

Seat No. _____

Room No. _____

School _____

Written Test: Weight 50, pass mark to be determined.

December 12, 1987

GENERAL INSTRUCTIONS FOR CANDIDATES—READ NOW

GENERAL INFORMATION: Read the General Information and the Test Instructions now and fill out the information requested at the top of this booklet. The following are explanations of the information requested.

The Social Security No. is your correct Social Security Number. If your Social Security Number is not correct on your admission card, fill out the pink correction form (DP-148) so the correction can be made after the test. Your Seat No. is the number shown on the sheet of scrap paper on your desk. Your Room No. and School are written on the blackboard.

THE FORMS: You must now fill out your fingerprint card and answer sheet as explained below. Use pencil only.

Fingerprint Card: Print in pencil all information requested including your Social Security Number. Your fingerprints will be taken approximately 15 minutes after the Seventh Signal.

Answer Sheet: Print in pencil all the information requested on the Answer Sheet including your Social Security No. and the Exam. No. in the boxes at the top of the columns, ONE NUMBER IN A BOX. In each column, darken (with pencil) the oval containing the number in the box at the top of the column; only **ONE OVAL** in each **COLUMN** should be darkened. Then, using your pencil, print the Exam, Title, School or Building, Room No., Seat No., and Today's Date (December 12, 1987) in the appropriate spaces.

After filling in this information at the top of the Answer Sheet, please read the Confidential Questionnaire at the bottom of the Answer Sheet and answer the questions accurately.

After answering the Confidential Questionnaire, please complete the Survey Questionnaire.

You are not permitted to use a calculator on this test.

TEST INSTRUCTIONS

EIGHT SIGNALS WILL BE USED DURING THIS TEST. The following is an explanation of these signals.

FIRST SIGNAL: The First Memory Booklet is distributed. Write your Room, Seat, and Social Security Numbers on the cover. DO NOT OPEN THE BOOKLET UNTIL THE SECOND SIGNAL.

SECOND SIGNAL: OPEN THE FIRST MEMORY BOOKLET. You will be given five minutes to remember as many details of the scene as you can. You may not write or make any notes while studying this material. The first eight questions will be based on this material.

THIRD SIGNAL: CLOSE THE FIRST MEMORY BOOKLET. The monitor will collect this booklet and distribute the Second Memory Booklet. Write your Room, Seat, and Social Security Numbers on top of the Second Memory Booklet. DO NOT OPEN THE SECOND MEMORY BOOKLET OR WRITE ANYTHING ELSE UNTIL THE NEXT SIGNAL.

FOURTH SIGNAL: OPEN THE SECOND MEMORY BOOKLET AND ANSWER QUESTIONS 1 TO 8. You will be given five minutes to answer the questions. DO NOT TURN TO ANY OTHER PART OF THE BOOKLET UNTIL THE NEXT SIGNAL. After you have answered questions 1 to 8 on your Answer Sheet, you may, for future reference, circle your answers in the Second Memory Booklet.

FIFTH SIGNAL: TURN TO THE DIAGRAM IN THE SECOND MEMORY BOOKLET. You will be given five minutes to try to remember as many details about the diagram as you can. You may not write or make any notes during this time. DO NOT TURN TO ANY OTHER PAGE DURING THIS TIME.

SIXTH SIGNAL: CLOSE THE SECOND MEMORY BOOKLET. Monitors will collect this booklet and distribute the Question Booklet. Do not open the Question Booklet until the next signal. DO NOT WRITE ANYTHING DURING THIS PERIOD.

SEVENTH SIGNAL: OPEN THE QUESTION BOOKLET AND ANSWER QUESTIONS 9 TO 16 DEALING WITH THE SECOND MEMORY SECTION. Then continue answering the rest of the questions in the Question Booklet. This test consists of 100 questions. After you finish the memory questions, check to make sure the Question Booklet has all the questions from 9 to 100 and is not defective. If you believe that your booklet is defective, inform the monitor. You will have 4 hours from this signal to complete all the questions. YOU WILL BE FINGERPRINTED DURING THIS PERIOD.

EIGHTH SIGNAL: END OF TEST. STOP WORKING. If you finish before this signal and want to leave, raise your hand.

ANYONE DISOBEYING ANY OF THESE INSTRUCTIONS MAY BE DISQUALIFIED—THAT IS, YOU MAY RECEIVE A SCORE OF ZERO FOR THE ENTIRE TEST!

DIRECTIONS FOR ANSWERING QUESTIONS	Answer all the questions on the Answer Sheet before the Eighth Signal is given. ONLY YOUR ANSWER SHEET WILL BE MARKED, although you also should circle your answers in the Question Booklet before the end of the test for future reference. Use a soft pencil (No. 2) to mark your answers. If you want to change an answer, erase it and then mark your new answer. For each question, pick the best answer. Then, on your Answer Sheet in the row with the same number as the question, blacken the oval with the same letter as your answer. Do not make any stray pencil dots, dashes, or marks any place on the Answer Sheet. Do NOT fold, roll, or tear the Answer Sheet.

Here is a sample of how to mark your answers:

SAMPLE 0: The sum of 5 and 3 is

(A) 11 (B) 9 (C) 8 (D) 2

Because the answer is 8, your Answer Sheet should be marked like this:

SAMPLE 0: (A) ○   (B) ○   (C) ●   (D) ○

WARNING:	You are not allowed to copy answers from anyone or to use books or notes. It is against the law to take the test for somebody else or to let somebody else take the test for you. There is to be NO SMOKING anywhere in the building.
LEAVING:	After the test starts, candidates may not leave the building until after they are fingerprinted. No one may leave the building until after 10:45 A.M. No one may come in after 10:45 A.M. During the test, you may not leave the room unless accompanied by a monitor. If you want to drop out of the test and not have your answers marked, write "I withdraw" on your Answer Sheet and sign your name.

You may take your Question Booklet and two Memory Booklets with you when you finish. The room monitor will return your Memory Booklets to you when you hand in your Answer Sheet.

Answer Sheet for Practice Examination 1

1. Ⓐ Ⓑ Ⓒ Ⓓ Ⓔ
2. Ⓐ Ⓑ Ⓒ Ⓓ Ⓔ
3. Ⓐ Ⓑ Ⓒ Ⓓ Ⓔ
4. Ⓐ Ⓑ Ⓒ Ⓓ Ⓔ
5. Ⓐ Ⓑ Ⓒ Ⓓ Ⓔ
6. Ⓐ Ⓑ Ⓒ Ⓓ Ⓔ
7. Ⓐ Ⓑ Ⓒ Ⓓ Ⓔ
8. Ⓐ Ⓑ Ⓒ Ⓓ Ⓔ
9. Ⓐ Ⓑ Ⓒ Ⓓ Ⓔ
10. Ⓐ Ⓑ Ⓒ Ⓓ Ⓔ
11. Ⓐ Ⓑ Ⓒ Ⓓ Ⓔ
12. Ⓐ Ⓑ Ⓒ Ⓓ Ⓔ
13. Ⓐ Ⓑ Ⓒ Ⓓ Ⓔ
14. Ⓐ Ⓑ Ⓒ Ⓓ Ⓔ
15. Ⓐ Ⓑ Ⓒ Ⓓ Ⓔ
16. Ⓐ Ⓑ Ⓒ Ⓓ Ⓔ
17. Ⓐ Ⓑ Ⓒ Ⓓ Ⓔ
18. Ⓐ Ⓑ Ⓒ Ⓓ Ⓔ
19. Ⓐ Ⓑ Ⓒ Ⓓ Ⓔ
20. Ⓐ Ⓑ Ⓒ Ⓓ Ⓔ
21. Ⓐ Ⓑ Ⓒ Ⓓ Ⓔ
22. Ⓐ Ⓑ Ⓒ Ⓓ Ⓔ
23. Ⓐ Ⓑ Ⓒ Ⓓ Ⓔ
24. Ⓐ Ⓑ Ⓒ Ⓓ Ⓔ
25. Ⓐ Ⓑ Ⓒ Ⓓ Ⓔ
26. Ⓐ Ⓑ Ⓒ Ⓓ Ⓔ
27. Ⓐ Ⓑ Ⓒ Ⓓ Ⓔ
28. Ⓐ Ⓑ Ⓒ Ⓓ Ⓔ
29. Ⓐ Ⓑ Ⓒ Ⓓ Ⓔ
30. Ⓐ Ⓑ Ⓒ Ⓓ Ⓔ
31. Ⓐ Ⓑ Ⓒ Ⓓ Ⓔ
32. Ⓐ Ⓑ Ⓒ Ⓓ Ⓔ
33. Ⓐ Ⓑ Ⓒ Ⓓ Ⓔ
34. Ⓐ Ⓑ Ⓒ Ⓓ Ⓔ
35. Ⓐ Ⓑ Ⓒ Ⓓ Ⓔ
36. Ⓐ Ⓑ Ⓒ Ⓓ Ⓔ
37. Ⓐ Ⓑ Ⓒ Ⓓ Ⓔ
38. Ⓐ Ⓑ Ⓒ Ⓓ Ⓔ
39. Ⓐ Ⓑ Ⓒ Ⓓ Ⓔ
40. Ⓐ Ⓑ Ⓒ Ⓓ Ⓔ
41. Ⓐ Ⓑ Ⓒ Ⓓ Ⓔ
42. Ⓐ Ⓑ Ⓒ Ⓓ Ⓔ
43. Ⓐ Ⓑ Ⓒ Ⓓ Ⓔ
44. Ⓐ Ⓑ Ⓒ Ⓓ Ⓔ
45. Ⓐ Ⓑ Ⓒ Ⓓ Ⓔ
46. Ⓐ Ⓑ Ⓒ Ⓓ Ⓔ
47. Ⓐ Ⓑ Ⓒ Ⓓ Ⓔ
48. Ⓐ Ⓑ Ⓒ Ⓓ Ⓔ
49. Ⓐ Ⓑ Ⓒ Ⓓ Ⓔ
50. Ⓐ Ⓑ Ⓒ Ⓓ Ⓔ
51. Ⓐ Ⓑ Ⓒ Ⓓ Ⓔ
52. Ⓐ Ⓑ Ⓒ Ⓓ Ⓔ
53. Ⓐ Ⓑ Ⓒ Ⓓ Ⓔ
54. Ⓐ Ⓑ Ⓒ Ⓓ Ⓔ
55. Ⓐ Ⓑ Ⓒ Ⓓ Ⓔ
56. Ⓐ Ⓑ Ⓒ Ⓓ Ⓔ
57. Ⓐ Ⓑ Ⓒ Ⓓ Ⓔ
58. Ⓐ Ⓑ Ⓒ Ⓓ Ⓔ
59. Ⓐ Ⓑ Ⓒ Ⓓ Ⓔ
60. Ⓐ Ⓑ Ⓒ Ⓓ Ⓔ
61. Ⓐ Ⓑ Ⓒ Ⓓ Ⓔ
62. Ⓐ Ⓑ Ⓒ Ⓓ Ⓔ
63. Ⓐ Ⓑ Ⓒ Ⓓ Ⓔ
64. Ⓐ Ⓑ Ⓒ Ⓓ Ⓔ
65. Ⓐ Ⓑ Ⓒ Ⓓ Ⓔ
66. Ⓐ Ⓑ Ⓒ Ⓓ Ⓔ
67. Ⓐ Ⓑ Ⓒ Ⓓ Ⓔ
68. Ⓐ Ⓑ Ⓒ Ⓓ Ⓔ
69. Ⓐ Ⓑ Ⓒ Ⓓ Ⓔ
70. Ⓐ Ⓑ Ⓒ Ⓓ Ⓔ
71. Ⓐ Ⓑ Ⓒ Ⓓ Ⓔ
72. Ⓐ Ⓑ Ⓒ Ⓓ Ⓔ
73. Ⓐ Ⓑ Ⓒ Ⓓ Ⓔ
74. Ⓐ Ⓑ Ⓒ Ⓓ Ⓔ
75. Ⓐ Ⓑ Ⓒ Ⓓ Ⓔ
76. Ⓐ Ⓑ Ⓒ Ⓓ Ⓔ
77. Ⓐ Ⓑ Ⓒ Ⓓ Ⓔ
78. Ⓐ Ⓑ Ⓒ Ⓓ Ⓔ
79. Ⓐ Ⓑ Ⓒ Ⓓ Ⓔ
80. Ⓐ Ⓑ Ⓒ Ⓓ Ⓔ
81. Ⓐ Ⓑ Ⓒ Ⓓ Ⓔ
82. Ⓐ Ⓑ Ⓒ Ⓓ Ⓔ
83. Ⓐ Ⓑ Ⓒ Ⓓ Ⓔ
84. Ⓐ Ⓑ Ⓒ Ⓓ Ⓔ
85. Ⓐ Ⓑ Ⓒ Ⓓ Ⓔ
86. Ⓐ Ⓑ Ⓒ Ⓓ Ⓔ
87. Ⓐ Ⓑ Ⓒ Ⓓ Ⓔ
88. Ⓐ Ⓑ Ⓒ Ⓓ Ⓔ
89. Ⓐ Ⓑ Ⓒ Ⓓ Ⓔ
90. Ⓐ Ⓑ � Ⓓ Ⓔ
91. Ⓐ Ⓑ Ⓒ Ⓓ Ⓔ
92. Ⓐ Ⓑ Ⓒ Ⓓ Ⓔ
93. Ⓐ Ⓑ Ⓒ Ⓓ Ⓔ
94. Ⓐ Ⓑ Ⓒ Ⓓ Ⓔ
95. Ⓐ Ⓑ Ⓒ Ⓓ Ⓔ
96. Ⓐ Ⓑ Ⓒ Ⓓ Ⓔ
97. Ⓐ Ⓑ Ⓒ Ⓓ Ⓔ
98. Ⓐ Ⓑ Ⓒ Ⓓ Ⓔ
99. Ⓐ Ⓑ Ⓒ Ⓓ Ⓔ
100. Ⓐ Ⓑ Ⓒ Ⓓ Ⓔ

TEAR HERE

DEPARTMENT OF PERSONNEL

Social Security No. _____
Room No. _____
Seat No. _____
School _____

First Memory Booklet　　**FIREFIGHTER**

December 12, 1987　　**EXAMINATION NO. 7022**

DO NOT OPEN THIS BOOKLET UNTIL THE SECOND SIGNAL IS GIVEN!

Write your Social Security Number, Room Number, Seat Number, and School in the appropriate spaces at the top of this page.

You must follow the instructions found in the CANDIDATES' INSTRUCTION BOOKLET.

ANYONE DISOBEYING ANY OF THE INSTRUCTIONS FOUND IN THE CANDIDATES' INSTRUCTION BOOKLET MAY BE DISQUALIFIED—RECEIVE A ZERO ON THE ENTIRE TEST.

This booklet contains one scene. Try to remember as many details in the scene as you can.

DO NOT WRITE OR MAKE ANY NOTES WHILE STUDYING THE SCENE.

DO NOT OPEN THIS BOOKLET UNTIL THE SECOND SIGNAL IS GIVEN!

　　Note: You will have five (5) minutes to study the scene shown.

Copyright 1987 by the Department of Personnel, City of New York. No part of this booklet may be reproduced by any method without the written permission of the Department of Personnel, City of New York.

　　The New York City Department of Personnel makes no commitment, and no inference is to be drawn, regarding the content, style, or format of any future examination for the position of firefighter.

NOTE: The ground floor is the first floor.

CITY OF NEW YORK
DEPARTMENT OF PERSONNEL

Social Security No. _____
Room No. _____
Seat No. _____
School _____

Second Memory Booklet   **FIREFIGHTER**

December 12, 1987   **EXAMINATION No. 7022**

DO NOT OPEN THIS BOOKLET UNTIL THE FOURTH SIGNAL IS GIVEN!

Write your Social Security Number, Room Number, Seat Number, and School in the appropriate spaces at the top of this page.

You must follow the instructions found in the CANDIDATES' INSTRUCTION BOOKLET.

ANYONE DISOBEYING ANY OF THE INSTRUCTIONS FOUND IN THE CANDIDATES' INSTRUCTION BOOKLET MAY BE DISQUALIFIED—RECEIVE A ZERO ON THE ENTIRE TEST.

This booklet contains questions 1 through 8, which are based on the scene shown in the First Memory Booklet. Answer questions 1 through 8 after the fourth signal.

At the FIFTH SIGNAL, you should turn to the Second Memory Diagram. You will have five (5) minutes to memorize the floor plan presented in the diagram.

DO NOT OPEN THIS BOOKLET UNTIL THE FOURTH SIGNAL IS GIVEN!

Copyright 1987 by the Department of Personnel, City of New York. No part of this booklet may be reproduced by any method without the written permission of the Department of Personnel, City of New York.

The New York City Department of Personnel makes no commitment, and no inference is to be drawn, regarding the content, style, or format of any future examination for the position of firefighter.

DIRECTIONS: *Answer questions 1 through 8 based on the scene in the first memory booklet.*

1. The fire is located on the
   - (A) first floor
   - (B) fourth floor
   - (C) fifth floor
   - (D) top floor

2. The smoke and flames are blowing
   - (A) up and to the left
   - (B) up and to the right
   - (C) down and to the left
   - (D) down and to the right

3. There is a person on a fire escape on the
   - (A) second floor
   - (B) third floor
   - (C) fourth floor
   - (D) fifth floor

4. People are visible in windows at the front of the building on fire on the
   - (A) second and third floors
   - (B) third and fifth floors
   - (C) fourth and sixth floors
   - (D) fifth and sixth floors

5. The person who is closest to the flames is in a
   - (A) front window on the third floor
   - (B) front window on the fifth floor
   - (C) side window on the fifth floor
   - (D) side window on the third floor

6. A firefighter is told to go to the roof of the building on fire. It would be correct to state that the firefighter can cross directly to the roof from
   - (A) the roof of the bank
   - (B) the roof of the factory
   - (C) either the bank or the factory
   - (D) neither the bank nor the factory

7. On which side of the building on fire are fire escapes visible?
   - (A) left   (B) front   (C) right   (D) rear

8. The hydrant on the sidewalk is
   - (A) in front of the bank
   - (B) between the bank and the apartments
   - (C) in front of the apartments
   - (D) between the apartments and the factory

DO NOT TURN PAGE UNTIL NEXT SIGNAL

DIRECTIONS: Look at this floor plan of an apartment. It is on the third floor of the building. The floor plan also indicates the public hallway.

CITY OF NEW YORK
DEPARTMENT OF PERSONNEL

Social Security No. _____
Room No. _____
Seat No. _____
School _____

Question Booklet

December 12, 1987

FIREFIGHTER

EXAMINATION NO. 7022

Written Test: Weight 50

Time allowed: 4 Hours

DO NOT OPEN THIS BOOKLET UNTIL THE SEVENTH SIGNAL IS GIVEN!

Record your answers on the Official Answer Sheet before the last signal. If you wish, you also may record your answers in the Question Booklet before the last signal is given. This will be your only record of answers.

This Question Booklet contains procedures and definitions that are to be used to answer certain questions. THESE PROCEDURES AND DEFINITIONS ARE NOT NECESSARILY THOSE OF THE NEW YORK CITY FIRE DEPARTMENT. HOWEVER, YOU ARE TO ANSWER THESE QUESTIONS SOLELY ON THE BASIS OF THE MATERIAL GIVEN.

After the seventh signal is given, open this Question Booklet and begin work. You will have 4 hours to complete this test.

YOU MAY WRITE IN THIS BOOKLET (INCLUDING THE DIAGRAMS) AND USE THE SCRAP PAPER ON YOUR DESK. If you need additional scrap paper, ask the monitor.

Remember, only your Official Answer Sheet will be rated, so be sure to mark all your answers on the Official Answer Sheet before the eighth signal. No additional time will be given for marking your answers after the test has ended.

DO NOT OPEN THIS BOOKLET UNTIL THE SEVENTH SIGNAL IS GIVEN!

Copyright 1987 by the Department of Personnel, City of New York. No part of this booklet may be reproduced by any method without the written permission of the Department of Personnel, City of New York.

The New York City Department of Personnel makes no commitment, and no inference is to be drawn, regarding the content, style, or format of any future examination for the position of firefighter.

DIRECTIONS: Answer questions 9 through 16 on the basis of the floor plan you have just studied.

9. Which room is farthest from the fire escape?
   (A) bedroom 2
   (B) bedroom 3
   (C) the kitchen
   (D) the dining room

10. Which one of the following rooms has only one door or doorway?
    (A) the living room
    (B) bedroom 1
    (C) the kitchen
    (D) the dining room

11. Which room can firefighters reach directly from the fire escape?
    (A) the dining room
    (B) the living room
    (C) bedroom 1
    (D) bedroom 3

12. Which room does not have a door or doorway leading directly to the foyer?
    (A) bathroom 1
    (B) bathroom 2
    (C) bedroom 1
    (D) the dining room

13. A firefighter leaving bathroom 2 would be in
    (A) bedroom 1
    (B) bedroom 2
    (C) bedroom 3
    (D) the foyer

14. Firefighters on the terrace would be able to enter directly into which rooms?
    (A) bedroom 1 and bathroom 1
    (B) bedroom 2 and bathroom 2
    (C) the dining room and the kitchen
    (D) the dining room and the living room

15. Which rooms have at least one window on two sides of the building?
    (A) bedroom 2 and the dining room
    (B) bedroom 2 and bedroom 3
    (C) the dining room and the living room
    (D) the dining room, bedroom 2, and bedroom 3

16. Firefighters can enter the kitchen directly from the foyer and
    (A) bedroom 1
    (B) the living room
    (C) bathroom 1
    (D) the dining room

Please check your Answer Sheet to make sure you have written in and blackened your Social Security Number correctly. If you have not yet written in and blackened your Social Security Number, please do so immediately. If you have incorrectly written in or blackened your Social Security Number, please correct the mistake immediately. Failure to correctly blacken your Social Security Number may result in your Answer Sheet NOT being rated.

17. When the door to a fire area is opened, all firefighters _____ positioned on one side of the entrance.
    (A) have been
    (B) should have
    (C) want to
    (D) should be
    (E) will have

18. Soon after the fire yesterday, it _____ necessary to turn off the sprinkler system to prevent further damage.
    (A) will be
    (B) would be
    (C) is
    (D) has been
    (E) was

19. Firefighter Curtis _____ up for promotion next week.
    (A) was
    (B) has
    (C) are
    (D) is
    (E) will

20. It is important that the first firefighters arriving at a fire scene act _____.
    (A) quickest
    (B) quicker
    (C) quickly
    (D) quick
    (E) quickliest

21. Firefighter McDaniel, _____ courage should be commended, saved the little boy's life.
    (A) whom
    (B) whose
    (C) that
    (D) who
    (E) which

DIRECTIONS: *Answer questions 22 through 24 based solely on the following passage.*

When there is a fire in a subway train, it may be necessary for firefighters to evacuate people from the trains by way of the tunnels. In every tunnel, emergency exit areas have stairways that can be used to evacuate people to the street from the track area. All emergency exits can be recognized by an exit sign near a group of five white lights.

There is a Blue Light Area located every 600 feet in the tunnel. These areas contain a power removal box, a telephone, and a fire extinguisher. Removal of power from the third rail is the first step firefighters must take when evacuating people through tunnels. When a firefighter uses the power removal box to turn off electrical power during evacuation procedures, the firefighter must immediately telephone the trainmaster and explain the reason for the power removal. Communication between the firefighter and the trainmaster is essential. If the trainmaster does not receive a phone call within four minutes after power removal, the power will be restored to the third rail.

22. When evacuating passengers through the subway tunnel, firefighters must *first*
    (A) telephone the trainmaster for assistance
    (B) remove electrical power from the third rail
    (C) locate the emergency exit in the tunnel
    (D) go to the group of five white lights

23. Immediately after using the power removal box to turn off the electrical power, a firefighter should
    (A) wait four minutes before calling the trainmaster
    (B) begin evacuating passengers through the tunnel
    (C) call the trainmaster and explain why the power was turned off
    (D) touch the third rail to see if the electrical power has been turned off

24. A group of five white lights in a subway tunnel indicates that
    (A) a telephone is available
    (B) the electrical power is off in the third rail
    (C) a fire extinguisher is available
    (D) an emergency exit is located there

DIRECTIONS: *Answer questions 25 and 26 based on the following passage.*

A new firefighter learns the following facts about his company's response area: All the factories are located between 9th Avenue and 12th Avenue from 42nd Street to 51st Street, all the apartment buildings are located between 7th Avenue and 9th Avenue from 47th Street to 51st Street, all the private houses are located between 5th Avenue and 9th Avenue from 42nd Street to 47th Street, and all the stores are located between 5th Avenue and 7th Avenue from 47th Street to 51st Street.

The firefighter also learns that the apartment buildings are all between four and six stories, the private houses are all between one and three stories, the factories are all between three and five stories, and the stores are all either one or two stories.

25. An alarm is received for a fire located on 8th Avenue between 46th Street and 47th Street. The firefighter should assume that the fire is in a
    (A) private house between 1 and 3 stories
    (B) private house between 4 and 6 stories
    (C) factory between 3 and 5 stories
    (D) factory between 4 and 6 stories

26. The company responds to a fire on 47th Street between 6th Avenue and 7th Avenue. The firefighter should assume that he will be responding to a fire in
    (A) a store of either 1 or 2 stories
    (B) a factory between 3 and 5 stories
    (C) an apartment building between 4 and 6 stories
    (D) a private house between 4 and 6 stories

DIRECTIONS: *Answer question 27 based on the following passage.*

During a recent day tour with an engine company, firefighter Sims was assigned to the control position on the hose. The company responded to the following alarms during this tour:

Alarm 1: At 9:30 AM, the company responded to a fire on the first floor of an apartment building. At the fire scene, firefighter Sims pulled the hose from the fire engine and assisted the driver in attaching the hose to the hydrant.

Alarm 2: At 11 AM, the company responded to a fire on the third floor of a vacant building. Firefighter Sims pulled the hose from the fire engine and went to the building on fire.

Alarm 3: At 1 PM, the company responded to a fire in a first-floor laundromat. Firefighter Sims pulled the hose from the fire engine and assisted the driver in attaching the hose to the hydrant.

Alarm 4: At 3 PM, the company responded to a fire on the fourth floor of an apartment building. Firefighter Sims pulled the hose from the fire engine and went to the building on fire.

Alarm 5: At 5:45 PM, the company responded to a fire on the second floor of a private house. Firefighter Sims pulled the hose from the fire engine and assisted the driver in attaching the hose to the hydrant.

27. The firefighter assigned to the control position assists the driver in attaching the hose to a hydrant when the fire is
    (A) in an apartment building
    (B) above the second floor
    (C) in a vacant building
    (D) below the third floor

DIRECTIONS: *Answer questions 28 through 30 based on the floor plan shown on page 172 and the following information.*

The floor plan represents a typical high-rise office building in midtown Manhattan. Numbers shown indicate room numbers. The pipe connections for the water supply system are outside the building at street level. Firefighters attach hoses to these connections to send water into the pipes in the building.

Questions 28 through 30 refer to a fire on the first floor in Room 111.

28. After fighting the fire in Room 111, firefighters are instructed to go immediately to the east-west hallway in the center of the building and search for victims in that hallway. Which one of the following lists all the rooms the firefighters should search?
    (A) 115, 117, 118, 119, 133, and 134
    (B) 125, 126, 127, 128, and 129
    (C) 107, 109, 125, 126, 127, and 128
    (D) 121, 122, 123, 124, 125, and 126

29. Firefighters are told to search Room 134. They enter the building from 40th Street. The shortest route for the firefighters to take to reach this room is
    (A) west in hallway E, north in hallway A, and then east in hallway C
    (B) west in hallway E, north in hallway A, east in hallway D, north in hallway B, and then west in hallway C
    (C) east in hallway E, north in hallway B, and then west in hallway C
    (D) east in hallway E, north in hallway B, west in hallway D, north in hallway A, and then east in hallway C

30. Firefighters in Truck 21 have been ordered to attach a hose to a connection outside the building. The firefighters cannot use connection A because 40th Street is blocked by traffic. What is the first connection to which the firefighters can drive?
    (A) Connection B   (B) Connection C   (C) Connection D   (D) Connection E

DIRECTIONS: *Answer questions 31 through 33 based on the following passage.*

Firefighters often know the appearance and construction features of apartments by recognizing general features on the outside of the building. The following are some general features of different types of buildings in New York City.

1. Old Law Tenements
   Height — 5 to 7 stories
   Width — 25 feet
   Fire Escapes — There will be a rear fire escape if there are two apartments per floor. There will be front and rear fire escapes if there are four apartments per floor.

2. Row Frames
   Height — 2 to 5 stories
   Width — 20 feet to 30 feet
   Fire Escapes — There will be a rear fire escape if the building is higher than 2 stories.

3. Brownstones
   Height — 3 to 5 stories
   Width — 20 feet to 25 feet
   Fire Escapes — If the Brownstone has been changed from a private home to a multiple dwelling, there will be a rear fire escape. Unchanged Brownstones have no fire escapes.

31. Upon arrival at a fire, a firefighter observes that the building is three stories high and 25 feet wide. There are fire escapes only in the rear of the building. The firefighter should conclude that the building is either a
    (A) row frame or an unchanged brownstone
    (B) row frame or an old law tenement with two apartments per floor
    (C) changed brownstone or an old law tenement with four apartments per floor
    (D) row frame or a changed brownstone

32. At another fire, the building is five stories high and 25 feet wide. There is a front fire escape. The firefighters should conclude that this building has
    (A) a rear fire escape because the building is a row frame higher than two stories
    (B) a rear fire escape because the building is an old law tenement with four apartments per floor
    (C) no rear fire escape because the building is a brownstone that has been changed into a multiple dwelling
    (D) no rear fire escape because the building has a front fire escape

33. At another fire, the building is four stories high and 30 feet wide. The building has no front fire escape. The firefighter should conclude that the building is
    (A) a row frame that has no rear fire escape
    (B) an old law tenement that has four apartments per floor
    (C) a row frame that has a rear fire escape
    (D) a brownstone that has been changed from a private home to a multiple dwelling

DIRECTIONS: *Answer questions 34 through 36 based on the following passage.*

Firefighters use two-way radios to alert other firefighters to dangerous conditions and to the need for help. Messages should begin with "MAY DAY" or "URGENT." "MAY DAY" messages have priority over "URGENT" messages. The following is a list of specific emergencies and the messages that should be sent.

"MAY DAY" Messages:

1. When a collapse is probable in the area where the firefighters are working: "MAY DAY—MAY DAY, collapse probable, GET OUT."

2. When a collapse has occurred in the area where the firefighters are working: "MAY DAY—MAY DAY, collapse occurred." The firefighter also should give the location of the collapse. If there are trapped victims, the number and condition of the trapped victims also should be given.

3. When a firefighter appears to be a heart attack victim: "MAY DAY—MAY DAY, CARDIAC." The location of the victim also should be given.

4. When anyone has a serious, life-threatening injury: "MAY DAY—MAY DAY." The firefighter also should describe the injury and should give the condition and the location of the victim.

"URGENT" Messages:

1. When anyone has a less serious injury that requires medical attention (for example, a broken arm): "URGENT—URGENT." The firefighter also should give the type of injury and the location of the victim.

2. When the firefighters should leave the building and fight the fire from the outside: "URGENT—URGENT, back out." The firefighter also should indicate the area to be evacuated.

3. "URGENT" messages also should be sent when firefighters' lives are endangered due to a drastic loss of water pressure in the hose.

34. Firefighters are ordered to extinguish a fire on the third floor of an apartment building. As the firefighters are operating the hose on the third floor, the stairway collapses and cuts the hose. What message should the firefighters send?
    (A) "URGENT—URGENT, back out."
    (B) "URGENT—URGENT, we have a loss of water on the third floor."
    (C) "MAY DAY—MAY DAY, collapse occurred on third floor stairway."
    (D) "MAY DAY—MAY DAY, collapse probable, GET OUT."

35. Two firefighters on the second floor of a vacant building are discussing the possibility of the floor's collapse. One of the firefighters clutches his chest and falls down. What message should the other firefighter send?
    (A) "MAY DAY—MAY DAY, firefighter collapse on the second floor."
    (B) "MAY DAY—MAY DAY, CARDIAC on the second floor."
    (C) "URGENT—URGENT, firefighter unconscious on the second floor."
    (D) "URGENT—URGENT, collapse probable on the second floor."

36. A firefighter has just decided that a collapse of the third floor is probable when he falls and breaks his wrist. What is the first message he should send?
    (A) "URGENT—URGENT, broken wrist on the third floor."
    (B) "MAY DAY—MAY DAY, broken wrist on the third floor."
    (C) "MAY DAY—MAY DAY, collapse probable, GET OUT."
    (D) "URGENT—URGENT, back out, third floor."

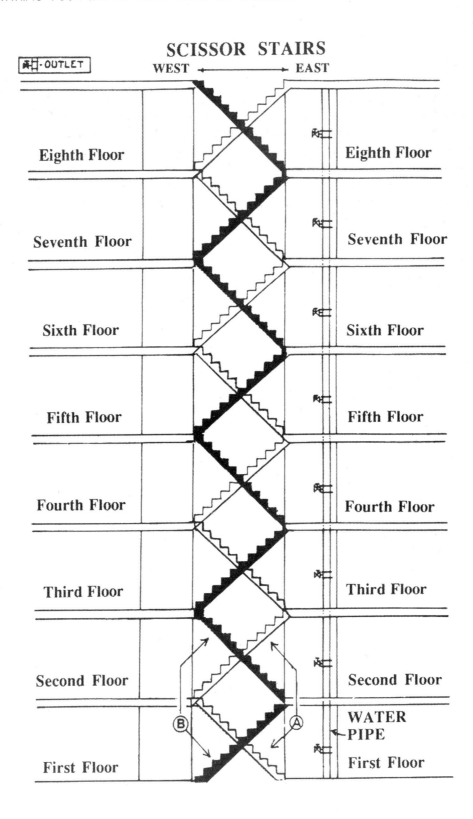

DIRECTIONS: *Answer questions 37 and 38 based on the diagram shown on page 176 and the following information.*

An eight-story apartment building has scissor stairs beginning on the first floor and going to the roof. Scissor stairs are two separate stairways (stairway A and stairway B) that crisscross each other and lead to opposite sides of the building on each floor. Once a person has entered either stairway, the only way to cross over to the other stairway on any floor is to leave the stairway and use the hallway on that floor. A person entering stairway A, which starts on the east side of the building on the first floor, would end up on the west side of the building on the second floor and back on the east side on the third floor. Similarly, a person entering stairway B, which starts on the west side of the building on the first floor, would end up on the east side of the building on the second floor and back on the west side on the third floor.

The apartment building has one water pipe for fighting fires. The pipe runs in a straight line near the stairway on the east side of the building from the first floor to the roof. There are water outlets for this pipe on each floor.

Questions 37 and 38 involve a fire in an apartment on the west side of the sixth floor.

37. Firefighters are ordered to connect a hose to the nearest outlet below the fire. Upon reaching this outlet, they find that it is not usable. The next available outlet is on the
    (A) fifth floor near stairway B
    (B) third floor near stairway A
    (C) fourth floor near stairway B
    (D) fourth floor near stairway A

38. A firefighter working on the west side of the seventh floor is ordered to search for victims on the west side of the eighth floor. The door leading to the stairway on the west side of the seventh floor is jammed shut. To reach the victims, the firefighter should take
    (A) stairway A to the eighth floor and then go across the hallway to the west side of the floor
    (B) stairway B to the eighth floor and then go across the hallway to the west side of the floor
    (C) the hallway to the east side of the seventh floor and go up stairway A
    (D) the hallway to the east side of the seventh floor and go up stairway B

39. Firefighters refer to the four sides of a building on fire as "exposures." The front of the fire building is referred to as Exposure 1. Exposures 2, 3, and 4 follow in clockwise order. Firefighters are working at a building with a front entrance that faces south. A firefighter in the center of the roof is ordered to go to Exposure 3. To reach Exposure 3, the direction in which he must walk is
    (A) east
    (B) west
    (C) south
    (D) north

DIRECTIONS: *Use the information in the following passage to answer questions 40 through 44.*

Ventilation is an important process involved in the extinguishing of a fire. A building on fire usually is ventilated by creating a hole in the roof or a wall of the structure. This allows heated air, smoke, and gases produced by a fire to be replaced with cool, fresh air. There are several advantages to the use of ventilation in firefighting including advantages in rescue operations, fire extinguishing, and property conservation.

Ventilation can make the rescue process faster and more efficient by removing gasses, smoke, and heated air that can be harmful to people who might be trapped in the building. By removing harmful substances produced by the fire and replacing them with fresh air, trapped individuals will have safer air to breathe and may have a better chance of survival. The removal of dark smoke and gases also can improve the visibility within the burning structure so that firefighters can locate victims faster.

The process of extinguishing a fire also can be improved by the proper use of ventilation. By creating a hole in the roof directly above a fire, gases, smoke, and heat will be effectively removed from the building. This will keep the fire from spreading to other parts of the building. At the same time, a properly placed hole will make conditions within the structure safer for firefighters by reducing heat within the building, removing dangerous gases, and improving visibility by removing smoke from the structure. This can help firefighters to quickly locate and extinguish the fire.

Property that is not damaged directly by the fire can be damaged by the heat and smoke produced by the fire as well as the water the firefighters use to extinguish the fire. This damage can be reduced, however, by the effective use of ventilation. By helping firefighters locate the source of a fire faster, ventilation can help confine the fire to a smaller area within the structure. This will reduce the amount of damaging heat and smoke while also reducing the amount of water necessary to extinguish the fire. With the fire confined to a small area, firefighters also can try to salvage some property by removing it from the structure even before the fire is completely extinguished.

40. Ventilation enables heated air, smoke, and gases to be replaced with
    (A) new water
    (B) victims
    (C) fire
    (D) fresh air
    (E) firefighters

41. Property that is not damaged directly by a fire can be damaged by heat and smoke as well as the _____ that the firefighters use.
    (A) chemicals
    (B) hoses
    (C) water
    (D) ventilation
    (E) axes

42. Improved visibility created by proper ventilation will improve the rescue process by helping firefighters locate
    (A) the water main
    (B) the smoke detector
    (C) the hose
    (D) the ceiling or wall
    (E) the victims

43. Creating a hole in the roof directly above the fire will keep the _____ from spreading to other parts of the building.
    (A) water
    (B) victims
    (C) firefighters
    (D) fire
    (E) ventilation

44. The preceding passage mentions three advantages to the use of ventilation in firefighting. These are advantages in
    (A) fire extinguishing, rescue operations, and fire control
    (B) fire extinguishing, rescue operations, and property conservation
    (C) rescue operations, property conservation, and wind control
    (D) rescue operations, property conservation, and fire conservation
    (E) property conservation, fire control, and rescue operations

DIRECTIONS: *Answer questions 45 through 50 based on the diagram on page 180 and the following information.*

At three o'clock in the morning, a fire alarm is received for the area shown in the diagram. A train loaded with highly flammable material is on fire. The entire area is surrounded by a 10-foot high fence. At the time of the fire, Gate A is open but Gates B, C, and D are locked.

45. The first engine company arrives at the fire scene. The security guard at Gate A informs the firefighters of the location of the fire. Firefighter Jensen knows the area. He should inform the lieutenant that the way to drive to a hydrant that is as close to the fire as possible without passing through the smoke and flames is by going
    (A) south on Main Gate Drive, east on Research Road, south on Dual Gate Drive, and west on Hi-Lo Lane to hydrant 3
    (B) south on Main Gate Drive, west on Warehouse Court, south on Factory Road, and west on Hi-Lo Lane to hydrant 4
    (C) south on Main Gate Drive and east on Research Road to hydrant 1
    (D) east on Hawthorne Street and south on Rutland Road to hydrant 7

46. Firefighters at Employee Parking Lot A are ordered to drive their truck to the fence outside Gate D. The shortest route the firefighters could take from Warehouse Court is
    (A) south on Factory Road, west on Hi-Lo Lane, and north on Trailer Street
    (B) east on Research Road and south on Dual Gate Drive
    (C) north on Main Gate Drive, east on Hawthorne Street, and south on Rutland Road
    (D) north on Main Gate Drive, west on Hawthorne Street, south on Trailer Street, and west on Hi-Lo Lane

47. The first ladder company arrives at the fire scene. As they are driving north on Rutland Road, firefighters see the fire through Gate D. They cut the locks and enter Gate D. The lieutenant orders a firefighter to go on foot from Gate D to the Research Building and to search it for occupants. The entrance to the Research Building that is closest to the firefighter is
    (A) connected to the Visitor Parking Lot
    (B) located on Research Road
    (C) connected to Parking Lot B
    (D) located on Dual Gate Drive

48. The second engine company to arrive is ordered to attach a hose to a hydrant located outside the fenced area and then to await further orders. The hydrant outside of the fenced area that is closest to the flames is
    (A) hydrant 6
    (B) hydrant 3
    (C) hydrant 4
    (D) hydrant 7

49. The second ladder company to arrive at the fire scene is met at Gate C by a security guard who gives them the keys to open all the gates. They drive south on Trailer Street to the corner of Hi-Lo Lane and Trailer Street. The company is then ordered to drive to the corner of Research Road and Dual Gate Drive. The shortest route for the company to take without being exposed to the smoke and flames is
    (A) east on Hi-Lo Lane, north on Factory Road, and east on Warehouse Court to Research Road
    (B) east on Hi-Lo Lane and north on Dual Gate Drive
    (C) north on Trailer Street, east on Hawthorne Street, and south on Dual Gate Drive
    (D) north on Trailer Street, east on Hawthorne Street, south on Main Gate Drive, and east on Research Road

50. The heat from the fire in the railroad cars ignites the warehouse on the other side of Hi-Lo Lane. The officer of the first ladder company orders two firefighters on the west end of the loading dock to break the windows on the north side of the warehouse. Of the following, the shortest way for the firefighters to reach the northwest corner of the warehouse without passing through the smoke and flames is to go
    (A) east on Hi-Lo Lane, north on Dual Gate Drive, and then west on Research Road to the entrance on Warehouse Court
    (B) west on Hi-Lo Lane, north on Factory Road, and then east on Warehouse Court to the Visitor Parking Lot on Warehouse Court
    (C) east on Hi-Lo Lane, north on Rutland Road, west on Hawthorne Street, and then south on Main Gate Drive to the Visitor Parking Lot on Warehouse Court
    (D) east on Hi-Lo Lane, north on Dual Gate Drive, west on Hawthorne Street, and then south on Main Gate Drive to the entrance on Warehouse Court

DIRECTIONS: *Answer question 51 based on the following passage.*

Firefighters might have to use tools to force open an entrance door. Before the firefighters use the tools, they should turn the doorknob to see if the door is unlocked. If the door is locked, one firefighter should use an ax and another firefighter should use the halligan tool. This tool is used to pry open doors and windows. Firefighters must take the following steps in the order given to force open the door:

1. Place the prying end of the halligan tool approximately six inches above or below the lock. If there are two locks, the halligan tool should be placed between them.

2. Tilt the halligan tool slightly downward so that a single point on the prying end is at the door's edge.

3. Strike the halligan tool with the ax until the first point is driven in between the door and the door frame.

4. Continue striking with the ax until the door and frame are spread apart and the lock is broken.

5. Apply pressure toward the door and the door will spring open.

51. Firefighters respond to a fire on the second floor of a three-story apartment building. Two firefighters, one equipped with an ax and the other with a halligan tool, climb the stairs to the apartment on fire. They see that there are two locks on the apartment door. The firefighters should now
    (A) place the prying end of the halligan tool about six inches above or below the locks
    (B) turn the doorknob to determine whether the door is locked
    (C) tilt the halligan tool towards the floor before striking
    (D) place the prying end of the halligan tool between the locks

DIRECTIONS: Answer questions 52 and 53 based on the following passage.

The firefighter assigned to the roof position at a fire in a brownstone building should perform the following steps in the order given:

1. Go to the roof using one of the following options:

    a. First choice—The aerial ladder.

    b. Second choice—An attached building of the same height as the fire building.

    c. Third choice—A rear fire escape.

    d. Fourth choice—A 35-foot portable ladder.

2. Upon arrival on the roof, look around to determine whether any people are trapped who cannot be seen from the street.

    a. If a trapped person is observed, notify the officer and the driver that a lifesaving rope rescue is required. While waiting for assistance to conduct this rescue, assure the victim that help is on the way and proceed to Step 3.

    b. If no trapped persons are visible, proceed directly to Step 3.

3. Remove the cover from the opening in the roof.

    a. If there is no smoke or very little smoke coming from the opening, report to the officer for further orders.

    b. If heavy smoke comes from the opening, proceed to Step 4.

4. Remove the glass from the skylight.

52. Firefighters arriving at a fire in a brownstone are using the aerial ladder to make an immediate rescue. The firefighter assigned to the roof position should go to the roof of the building on fire via
    (A) a 35-foot portable ladder
    (B) a rear fire escape
    (C) an attached building of the same height
    (D) the inside stairway of the fire building

53. The firefighter assigned to the roof position at a fire in a brownstone arrives at the roof and finds that no persons are trapped. He then removes the roof cover from the opening in the roof. Which one of the following steps should be performed next?
    (A) He should remove the glass from the skylight if heavy smoke is coming from the opening.
    (B) He should remove the glass from the skylight if no smoke is coming from the roof opening.
    (C) He should go to the top floor to assist in the search for trapped persons if heavy smoke is coming from the roof opening.
    (D) He should report to the officer if heavy smoke is coming from the roof opening.

DIRECTIONS: *Answer questions 54 and 55 based on the following passage.*

Firefighters often are required to remove people who are trapped in elevators. At this type of emergency, firefighters perform the following steps in the order given:

1. Upon entering the building, determine the location of the elevator involved.
2. Reassure the trapped occupants that the fire department is on the scene and that firefighters are attempting to free them.
3. Determine whether there are any injured people in the elevator.
4. Determine whether all the doors from the hallways into the elevator shaft are closed.
5. If all the doors are closed, call for an elevator mechanic.
6. Wait until a trained elevator mechanic arrives before attempting to remove any trapped persons from the elevator unless they can be removed through the door to the hallway. Firefighters must remove the trapped persons by any safe method, however, if any one of the following conditions exists:

    a. There is a fire in the building.
    b. Someone in the elevator is injured.
    c. The people trapped in the elevator are in a state of panic.

54. Firefighters arrive at an elevator emergency in an office building. When they arrive, a maintenance man directs them to an elevator stuck between the fourth and fifth floors. He informs the firefighters that there is a young man in the elevator who apparently is calm and unhurt. Which one of the following steps should the firefighters perform next?
    (A) Determine whether the young man is injured.
    (B) Reassure the young man that the fire department is on the scene and that firefighters are attempting to free him.
    (C) Check to make sure all the doors to the elevator and hallways are closed.
    (D) Call for an elevator mechanic and await his arrival.

55. Firefighters are called to an elevator emergency at a factory building. The freight elevator has stopped suddenly between floors. The sudden stop caused heavy boxes to fall on the elevator

operator, breaking his arm. Upon arrival, the firefighters determine the location of the elevator. They tell the trapped operator that they are on the scene, are aware of his injury, and are attempting to free him. They determine that all the hallway doors leading into the elevator shaft are closed. The firefighters' next step should be to

(A) call for an ambulance and wait until it arrives
(B) remove the trapped person through the door to the hallway
(C) call for an elevator mechanic
(D) remove the trapped person by any safe method

DIRECTIONS: *Answer question 56 based on the following passage.*

The preferred order of actions for firefighters to take when removing a victim from an apartment on fire is as follows:

1. First choice—Remove the victim to the street level through the public hallway.

2. Second choice—Remove the victim to the street level by the fire escape.

3. Third choice—Remove the victim to the street level using either a portable ladder or an aerial ladder.

4. Fourth choice—Lower the victim to the street level with a lifesaving rope.

56. Firefighters answering an alarm are not able to use the entrance to a building on fire to reach a victim on the third floor because there is a fire in the public hallway. The victim is standing at the front window of an apartment on fire that has a fire escape. A firefighter places a portable ladder against the building, climbs the ladder, and enters the window where the victim is standing. The firefighter is carrying a lifesaving rope and a two-way radio. The radio enables her to communicate with the firefighter who operates the aerial ladder. The firefighter should then remove this victim from the apartment by using the

(A) fire escape
(B) aerial ladder
(C) portable ladder
(D) lifesaving rope

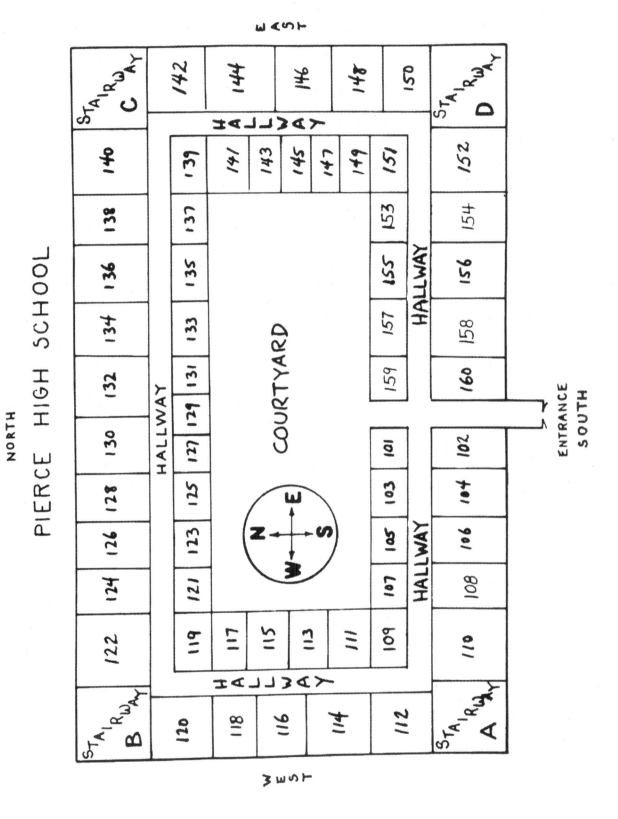

DIRECTIONS: *Answer questions 57 through 61 based on the floor plan on page 186 and the following information.*

Because many children might need to be rescued in the event of a school fire, New York City firefighters must become familiar with the floor layouts of public schools. Firefighters can develop this familiarity by conducting training drills at the schools.

A ladder company and an engine company recently conducted a drill at Pierce High School. The firefighters determined that the room layout is the same on all floors.

Several days after the drill, the ladder and engine companies report to a fire at Pierce High School in classroom 304, which is on the third floor. The fire has spread into the hallway in front of Room 304, blocking the hallway.

57. A firefighter is instructed to search for victims in the southwest area of the third floor. He wants to search as many rooms as possible and start his search as close to the fire as possible without passing through the fire. From the street, the firefighter should use his ladder to enter
   (A) Room 302
   (B) Room 306
   (C) Room 312
   (D) Room 352

58. A firefighter goes to the third floor by way of the southwest building stairway. In Room 317, he finds a child who has been overcome by smoke. Upon returning to the hallway, he finds that the stairway he came up is now blocked by fire hoses. The closest stairway the firefighter can use to bring the child to the street level is
   (A) Stairway A
   (B) Stairway B
   (C) Stairway C
   (D) Stairway D

59. Fire has spread from Room 304 to the room directly across the hall. As a result of heavy smoke, firefighters are ordered to break the windows of this room from the closest room on the floor above. Which room should the firefighters go to?
   (A) Room 403   (B) Room 305   (C) Room 313   (D) Room 413

60. The fire is in Rooms 303 and 305. Firefighters are told to go to the rooms in the north corridor facing the courtyard that are directly opposite 303 and 305. Which rooms should the firefighters go to?
   (A) Rooms 323 and 325
   (B) Rooms 326 and 328
   (C) Rooms 333 and 335
   (D) Rooms 355 and 357

61. Another fire breaks out in Room 336, blocking the entire hallway. Firefighters have brought a hose up the northeast stairway to fight this fire. Another hose must be brought up another stairway so that firefighters can approach the fire from the same direction. What is the closest stairway the firefighters could use?
   (A) Stairway A   (B) Stairway B   (C) Stairway C   (D) Stairway D

62. When attempting to rescue a person trapped at a window, firefighters frequently use an aerial ladder with a rotating base. The rotating base enables the ladder to move up and down and from side to side. In a rescue situation, the ladder should be placed so that both sides of the ladder rest fully on the window sill after it has been raised into position. Which one of the following diagrams shows the proper placement of the rotating base in relation to the window where the person is trapped?

DIRECTIONS: *Answer question 63 based solely on the following passage.*

Firefighters must regularly inspect office buildings to determine whether fire prevention laws have been obeyed. Some of these fire prevention laws are as follows:

DOORS: Doors should be locked as follows:

1. Doors on the ground floor may be locked on the street side to prevent entry into the stairway.

2. Doors in office buildings that are less than 100 feet in height may be locked on the stairway side on each floor above the ground floor.

3. Doors in office buildings that are 100 feet or more in height may be locked on the stairway side except for every fourth floor.

63. The doors in an office building that is less than 100 feet in height may be locked on the stairway side
    (A) on all floors including the ground floor
    (B) on all floors above the ground floor
    (C) except for every fourth floor
    (D) on all floors above the fourth floor

DIRECTIONS: *Answer question 64 based solely on the following passage.*

SIGNS: Signs concerning stairways should be posted in the following manner:

1. A sign shall be posted near the elevator on each floor, stating "IN CASE OF FIRE, USE STAIRS UNLESS OTHERWISE INSTRUCTED." The sign shall contain a diagram showing the location of the stairs and the letter identification of the stairs.

2. Each stairway shall be identified by an alphabetical letter on a sign posted on the hallway side of the stair door.

3. Signs indicating the floor number shall be attached to the stairway side of each door.

4. Signs indicating whether re-entry can be made into the building (and the floors where re-entry can be made) shall be posted on the stairway side of each door.

64. Which one of the following correctly lists the information that should be posted on the stairway side of a door?
    (A) A sign should indicate the floor number, whether re-entry can be made into the building, and the floors where re-entry can be made.
    (B) A sign should indicate the alphabetical letter of the stairway, whether re-entry can be made into the building, and the floors where re-entry can be made.
    (C) A sign should indicate the alphabetical letter of the stairway and the floor number.
    (D) A sign should indicate the alphabetical letter of the stairway, the floor number, whether re-entry can be made into the building, and the floors where re-entry can be made.

DIRECTIONS: *Answer question 65 based solely on the following passage.*

Every firefighter must know the proper first-aid procedures for treating injured people when there is a subway fire. In general, anyone suffering from smoke inhalation or heat exhaustion should be removed to fresh air and given oxygen immediately. Heart attack victims should be kept calm and should receive oxygen and medical attention immediately. Persons suffering from broken bones should not be moved until a splint is applied to the injury. In a situation in which there is a smoky fire in the subway, however, and the passengers needing immediate first aid are in danger from the fire, firefighters must first evacuate the passengers and perform first aid later, regardless of the injury.

65. The proper first-aid procedure for a man who apparently has suffered a heart attack on the station platform is to
    (A) have the man take the next train to the nearest hospital
    (B) remove the man to the street and administer oxygen
    (C) turn off the electrical power and evacuate the man through the tunnel
    (D) keep the man calm and administer oxygen

DIRECTIONS: *Answer questions 66 and 67 based solely on the following passage.*

A firefighter is responsible for a variety of duties other than fighting fires. One such duty is housewatch.

A firefighter's primary responsibility during housewatch is to properly receive alarm information. This enables firefighters to respond to alarms for fires and emergencies. The alarms are received at the firehouse by one of the following methods: computer teleprinter messages, fire department telephone, or verbal alarm. The computer teleprinter and the telephone are used to alert the fire companies. These two types of alarms are transmitted by a dispatcher from a central communication office to the firehouse closest to the fire. The verbal alarm occurs when someone comes to the firehouse or stops the fire truck on the street to report a fire. Once an alarm has been received, the firefighter on housewatch duty alerts the rest of the firefighters to respond to the alarm.

Other housewatch responsibilities include keeping the appearance of the housewatch area neat and orderly, keeping the front of the firehouse clear of all vehicles and obstructions, and receiving telephone calls and visitors with complaints about fire hazards. The firefighter on housewatch duty also keeps an accurate and complete record of all administrative matters in a journal.

66. The methods a dispatcher uses to transmit alarms to the firehouse are the
    (A) computer teleprinter, fire department telephone, and verbal alarm
    (B) verbal alarm and computer teleprinter
    (C) Fire department telephone and verbal alarm
    (D) computer teleprinter and fire department telephone

67. The primary responsibility of a firefighter on housewatch duty is to
   (A) properly assign firefighters to specific duties
   (B) properly receive alarm information
   (C) keep the housewatch area neat and orderly
   (D) write all important information in the company journal

DIRECTIONS: *Answer questions 68 and 69 based solely on the following passage.*

One duty of a firefighter on housewatch is to ensure that the computer teleprinter is working properly. A company officer should be notified immediately of any equipment problems. The firefighter on housewatch should check the amount of paper in the teleprinter and should refill it when necessary. The firefighter also should check the selector panel on the computer. The selector panel has a series of buttons that are used by the firefighter to let the dispatcher know that an alarm has been received and that the fire company is responding. These buttons have lights. To make sure the computer is functioning properly, the firefighter should press the button marked "test" and then release the button. If the computer lights go on and then go off after the "test" button has been released, the computer is working properly. In addition, the light next to the "test" button should always be blinking.

68. To make sure the selector panel of the computer is working properly, the firefighter on housewatch duty presses the button marked "test" and then releases the button. The firefighter should conclude that the computer is working properly if the
   (A) computer lights stay on
   (B) computer lights keep blinking
   (C) computer lights go on and then off
   (D) "test" light stays on

69. A firefighter on housewatch duty notices that the teleprinter is almost out of paper. In this situation, the firefighter should
   (A) test the computer panel by pushing the "test" button
   (B) notify the officer to replace the paper
   (C) place a new supply of paper in the teleprinter
   (D) notify the dispatcher that the paper is being changed

DIRECTIONS: *Use the information in the following passage to answer questions 70 and 71.*

When Sara was coming home from work one day, she noticed a strange man leaving through the back door of her neighbor's house. She saw him from a distance, but it appeared that the man was white, in his mid 20s, 5' 11", 180 lbs., had long brown hair, and was wearing blue jeans, a green sweatshirt, and a red baseball cap. About 20 minutes later, she noticed smoke coming from her neighbor's house and immediately called the fire department. When the firefighters arrived, she told them about the man she saw leaving the house. Later that evening, she was told that they suspected the fire was caused by arson. She was asked to indicate if the man she had seen leaving her neighbor's house was among the suspects. The following suspects were in the lineup:

1. Male, white, late 20s, 6', 150 lbs., long brown hair, wearing blue jeans, a blue T-shirt, and a red baseball cap.

2. Male, white, mid 20s, 5' 11", 200 lbs., long blond hair, wearing blue jeans, a green sweatshirt, and a red baseball cap.

3. Male, white, mid 20s, 5' 11", 190 lbs., long brown hair, wearing blue jeans, a green sweatshirt, and a red baseball cap.

4. Male, white, mid 20s, 6' 2", 190 lbs., long brown hair, wearing blue jeans, a blue sweatshirt, and a red baseball cap.

5. Male, white, mid 30s, 6', 180 lbs., long brown hair, wearing blue jeans, a blue sweatshirt, and a red baseball cap.

70. Which one of the suspects in the lineup is most likely the man Sara saw leaving her neighbor's house?
    (A) #1
    (B) #2
    (C) #3
    (D) #4
    (E) #5

71. Which suspect can be ruled out as the man leaving the house because of his hair and weight?
    (A) #1
    (B) #2
    (C) #3
    (D) #4
    (E) #5

DIRECTIONS: *Answer question 72 based on the following passage.*

Firefighters from the first arriving ladder company work in teams while fighting fires in private homes. The inside team enters the building through the first-floor entrance and then searches the first floor for victims. The outside team uses ladders to enter upper-level windows for a quick search of the bedrooms on the second floor and above. The assignments for the members of the outside team are as follows:

Roof person—This member places a ladder at the front porch and enters the second floor windows from the roof of the porch.

Outside vent person and driver—These members work together and place a portable ladder at a window on the opposite side of the house from which the roof person is working. If the aerial ladder can be used, however, the outside vent person and driver climb the aerial ladder in the front of the house, and the roof person places a portable ladder on the left side of the house.

To search all four sides of a private home on the upper levels, firefighters from the second arriving ladder company place portable ladders at the sides of the house not covered by the first ladder company and enter the home through the upper-level windows.

72. The second ladder company to arrive at a fire in a two-story private home sees the aerial ladder being raised to the front porch roof. In this situation, the firefighters should place their portable ladders
  (A) to the left and right sides of the house because there is a front porch
  (B) to the rear and right sides of the house because the aerial ladder is being used
  (C) to the rear and left sides of the house because there is a front porch
  (D) to the left and right sides of the house because the aerial ladder is being used

DIRECTIONS: *Answer question 73 based on the following passage.*

The priority for the removal of a particular victim by aerial ladder depends on the following conditions:

If two victims are at the same window and are not seriously endangered by spreading fire, the victim who is easier to remove is taken down the ladder first and is helped safely to the street. In general, the term "easier to remove" refers to the victim who is more capable of being moved and more able to cooperate. After the easier removal is completed, time can be spent on the more difficult removal.

If there are victims at two different windows, the aerial ladder is first placed to remove the victims who are the most seriously endangered by the fire. The ladder is then placed to remove the victims who are less seriously exposed to the fire.

73. Assume that you are working at a fire and that there are a total of three victims at two windows. Victims 1 and 2 are at the same window, which is three floors above the fire and shows no evidence of heat or smoke. Victim 1 is a disabled, 23-year-old male, and Victim 2 is a 40-year-old woman. Victim 3, a 16-year-old male, is at a window of the apartment on fire. From your position in the street, you can see heavy smoke coming from this window and flames coming out of the window next to it. Which one of the following is the proper order for victim removal?
  (A) Victim 3, Victim 2, Victim 1
  (B) Victim 1, Victim 2, Victim 3
  (C) Victim 1, Victim 3, Victim 2
  (D) Victim 3, Victim 1, Victim 2

DIRECTIONS: *Answer questions 74 and 75 based on the following passage.*

The four different types of building collapses are as follows:

1. *Building Wall Collapse*—An outside wall of the building collapses, but the floors maintain their positions.

2. *Lean-To Collapse*—One end of the floor collapses onto the floor below it. This leaves a sheltered area on the floor below.

3. *Floor Collapse*—An entire floor falls to the floor below it, but large pieces of machinery in the floor below provide spaces that can provide shelter.

4. *Pancake Collapse*—A floor collapses completely onto the floor below it, leaving no spaces. In some cases, the force of this collapse causes successive lower floors to collapse.

74. The most serious injuries are likely to occur in
    (A) pancake collapses
    (B) lean-to collapses
    (C) floor collapses
    (D) building wall collapses

75. Of the following, a floor collapse is most likely to occur in a
    (A) factory building
    (B) private home
    (C) apartment building
    (D) hotel

DIRECTIONS: Answer question 76 based solely on the following passage.

Many subway tunnels contain a set of three rails used for train movement. The subway trains run on two rails. The third rail carries electricity and is the source of power for all trains. Electricity travels from the third rail through metal plates called contact shoes, which are located near the wheels on every train car. Electricity then travels through the contact shoes into the train's motor. Firefighters must be very careful when operating near the third rail because contact with the third rail can result in electrocution.

76. The source of power for subway trains is the
    (A) third rail
    (B) contact shoes
    (C) motor
    (D) metal plates

DIRECTIONS: Answer questions 77 through 81 based solely on the following passage.

Firefighters receive an alarm for an apartment fire on the fourth floor of a 14-story housing project at 1191 Park Place. One firefighter shouts the address as the other firefighters are getting on the fire truck. Knowledge of the address helps the firefighters decide which equipment to pull off the fire truck when they reach the fire scene.

The firefighters know where the water outlets are located in the building on fire. There is an outlet in every hallway. Firefighters always attach the hose at the closest outlet on the floor below the fire.

As they arrive at 1191 Park Place, three firefighters immediately take one length of hose each and go into the building. Because an officer has been told by the dispatcher that two children are trapped in the rear bedroom, the officer and two firefighters begin searching for victims and opening windows immediately upon entering the apartment on fire.

As in all housing project fires, the roof person goes to the apartment above the apartment on fire. From this position, he attaches a tool to a rope in order to break open the windows of the apartment on fire. From this position, the roof person also could make a rope rescue of a victim in the apartment on fire.

77. A firefighter shouted the address of the fire when the alarm was received so that the firefighters would
    (A) know which equipment to take from the truck at the fire scene
    (B) be more alert when they arrived at the fire
    (C) be prepared to make a rope rescue
    (D) know that two children were reported to be trapped

78. The hose should be attached to an outlet on the
    (A) floor above the fire
    (B) ground floor
    (C) fire floor
    (D) floor below the fire

79. Because of the information given to the officer by the dispatcher, the officer and two firefighters
    (A) entered the apartment above the fire for a rope rescue
    (B) began immediately to search for victims and to open windows
    (C) opened all the windows before the hose was moved in
    (D) attached a hose and moved to the origin of the fire

80. The roof person broke the windows of the apartment on fire with
    (A) an ax while leaning out the windows
    (B) an ax attached to the end of a rope
    (C) a tool while standing on the roof
    (D) a tool attached to a rope

81. The proper location for a rescue by the roof person at a fire in a housing project is the
    (A) hallway
    (B) apartment below the apartment on fire
    (C) apartment above the apartment on fire
    (D) fire escape

DIRECTIONS: *Answer question 82 based on the following passage.*

You are a firefighter inspecting a building for violations. You must perform the following steps in the order given:

1. Find the manager of the building and introduce yourself.

2. Have the manager accompany you during the inspection.

3. Start the inspection by checking the fire department permits that have been issued to the building. The permits are located in the office of the building.

4. Inspect the building for violations of the Fire Prevention Laws. Begin at the roof and work down to the basement or cellar.

5. As you inspect, write on a piece of paper any violations you find and explain them to the building manager.

82. You are inspecting a supermarket. After entering the building, you identify yourself to the store manager and ask him to come along during the inspection. Which one of the following actions should you take next?
   (A) Start inspecting the supermarket, beginning at the basement.
   (B) Start inspecting the supermarket, beginning at the roof.
   (C) Ask to see the fire department permits that have been issued to the supermarket.
   (D) Write down any violations you notice while introducing yourself to the manager.

DIRECTIONS: *Use the following information and table to answer questions 83 through 85.*

For a fuel to burn while it is in a gaseous state, it must mix with oxygen. This oxygen usually is found in the air. If there is too much or too little fuel in the mixture for the amount of oxygen, however, the fuel will not burn.

FLAMMABLE RANGES OF COMMON FUELS

Fuel	Lower Limit (%)	Upper Limit (%)
Acetylene	2.5	100
Hydrogen	4.0	75
Propane	2.2	9.5
Natural Gas	5.0	15
Gasoline	1.4	7.6

83. When mixed with oxygen, what percentage of propane would be below what is necessary for the mixture to burn?
   (A) 1.4%
   (B) 2.2%
   (C) 2.5%
   (D) 4.0%
   (E) 5.0%

84. When mixed with oxygen, what percentage of hydrogen would be above what is necessary for the mixture to burn?
   (A) 75%
   (B) 9.5%
   (C) 7.6%
   (D) 15%
   (E) 100%

85. When mixed with oxygen, what percentage of natural gas would be acceptable for the mixture to burn?
   (A) 30%
   (B) 2.2%
   (C) 10%
   (D) 2.5%
   (E) 25%

DIRECTIONS: Answer question 86 based on the following passage.

When administering first aid to a person who is severely bleeding, firefighters should perform the following steps in the order given:

1. Direct pressure

    (a) Press a dressing directly on the wound.

2. Elevation

    a. Lift the injured body part above the heart level while continuing direct pressure. This slows down the movement of blood to the wound.

    b. Never elevate a body part if it might be fractured.

    c. If the bleeding continues while the injured body part is elevated and is receiving direct pressure, pressure should be applied to an artery or another pressure point.

86. A firefighter arrives at an automobile accident and finds a woman who is bleeding severely from a cut just above the ankle. She also has a fracture of the upper arm. After placing a dressing on the ankle wound, the firefighter next should
    (A) apply pressure to a pressure point to stop the bleeding
    (B) elevate the upper arm and apply direct pressure to it
    (C) apply direct pressure to an artery in the arm
    (D) apply direct pressure to the leg while elevating it

DIRECTIONS: Answer question 87 based solely on the following passage.

One firefighter of the first company to arrive at a fire in a private house is assigned to the roof position. During the beginning stages of a fire, the roof person is part of a team that enters and searches the building for victims.

The roof person's duties are performed in the following order:

1. Climb a portable ladder to the front porch roof, break a window, and enter the building through this opening if there is no victim in immediate danger visible at another window. If there is such a victim, place the ladder at that window.

2. If there is no front porch, use the portable extension ladder at a side of the house.

3. Enter the house after breaking a window.

87. The roof person at a fire in a private house sees a boy at a top floor window on the right side of the house. There is a porch in the front of the house. The roof person should now place the portable ladder at the
    (A) front of the porch
    (B) top floor front window
    (C) top floor window on the right side of the house
    (D) top floor window on the left side of the house

DIRECTIONS: *For questions 88 and 89, identify which one of the underlined words is spelled incorrectly.*

88. The <u>feasibility</u> of opening a floor for <u>ventelation</u> during a firefighting <u>operation</u> <u>obviously</u> depends on how the floor was <u>constructed</u> and with what material.
    - (A) feasibility
    - (B) ventelation
    - (C) operation
    - (D) obviously
    - (E) constructed

89. Due to a <u>suceptibility</u> to <u>deterioration</u> and wear, fire hoses must be <u>maintained</u> and <u>thoroughly</u> <u>inspected</u> regularly.
    - (A) suceptibility
    - (B) deterioration
    - (C) maintained
    - (D) thoroughly
    - (E) inspected

90. The decimal .20 is equivalent to what fraction?
    - (A) $\frac{1}{4}$
    - (B) $\frac{1}{5}$
    - (C) $\frac{1}{6}$
    - (D) $\frac{1}{7}$
    - (E) $\frac{1}{8}$

91. If a fire engine is 8 feet tall and the ladder attached to the top of the engine has three sections that are 18 feet each in length, how high will the ladder reach if all three sections are fully extended?
    - (A) 52 feet
    - (B) 56 feet
    - (C) 62 feet
    - (D) 65 feet
    - (E) 68 feet

92. If for every mile per hour in speed that a fire engine is traveling it takes $\frac{3}{4}$ of a foot to come to a complete stop, how many feet will it take a fire engine traveling 60 mph to come to a complete stop?
    - (A) 60 feet
    - (B) 50 feet
    - (C) 45 feet
    - (D) 30 feet
    - (E) 65 feet

93. If water travels through a fire hose at a rate of 5 feet per second, how long will it take a hose that is 125 feet long to fill with water?
    - (A) 50 seconds
    - (B) 35 seconds
    - (C) 25 seconds
    - (D) 12 seconds
    - (E) 10 seconds

94. Fire Station 35 received a call at 4:20 PM, and the firefighters arrived at the scene at 4:25 PM. The fire was extinguished at 5:15 PM, and the firefighters arrived back at the station at 5:30 PM. How much time (in minutes) elapsed from the time of the call to the firefighters arrival back at the station?
    - (A) 60 minutes
    - (B) 70 minutes
    - (C) 75 minutes
    - (D) 80 minutes
    - (E) 90 minutes

DIRECTIONS: Use the following chart to answer questions 95 through 97.

STATION CALLS BY TYPE AND YEAR

	1990	1991	1992	1993
House Fire	87		68	73
Car Accident	154	167	174	186
Industrial Fire	56	48	59	43
Total	297	312	301	302

95. How many house-fire calls were received in 1991?
    (A) 83   (B) 97   (C) 73   (D) 104   (E) 86

96. What was the percent increase in car accidents from 1990 to 1993?
    (A) 21%   (B) 22%   (C) 25%   (D) 26%   (E) 28%

97. What was the average number of industrial fires from 1991 to 1993?
    (A) 47   (B) 48   (C) 49   (D) 50   (E) 51

98. A firefighter assigned to the roof position at a fire must notify the officer of dangerous conditions that can be seen from the roof. Which one of the following conditions is the most dangerous?
    (A) The roof is sagging and might collapse because of the fire on the top floor.
    (B) Rubbish is visible on the roof of the building next door.
    (C) The stairway to the roof of the building has poor lighting.
    (D) An automobile accident in the street is causing a traffic jam.

99. Firefighters must often deal with people who need medical assistance. In life-threatening situations, firefighters must perform first aid until an ambulance arrives. In less serious situations, firefighters should make the person comfortable and should wait for the ambulance personnel to give first aid. A firefighter should give first aid until an ambulance arrives when a person
    (A) appears to have a knee injury
    (B) is bleeding heavily from a stomach wound
    (C) has bruises on his head
    (D) has a broken ankle

100. At a fire on the west side of the third floor of a 10-story office building, a firefighter is responsible for rescuing trapped persons by means of the aerial ladder. The trapped person in the most dangerous location should be removed first. Of the following, the firefighter should *first* remove the trapped person
    (A) on the rear fire escape on the east side of the second floor
    (B) on the roof
    (C) at a window on the west side of the fourth floor
    (D) at a window on the tenth floor

Answer Key for Practice Examination 1

Proposed Final Key Answers for Written Test Held on December 12, 1987.

1. B	26. A	51. B	76. A
2. B	27. D	52. C	77. A
3. D	28. C	53. A	78. D
4. B	29. C	54. B	79. B
5. B	30. D	55. C	80. D
6. D	31. D	56. A	81. C
7. A	32. B	57. B	82. C
8. A	33. C	58. B	83. A
9. D	34. C	59. A	84. E
10. B	35. B	60. A	85. C
11. D	36. C	61. D	86. D
12. B	37. C	62. A	87. **DELETED**
13. B	38. C	63. B	88. B
14. D	39. D	64. A	89. A
15. A	40. D	65. D	90. B
16. D	41. C	66. D	91. C
17. D	42. E	67. B	92. C
18. E	43. D	68. C	93. C
19. D	44. B	69. C	94. B
20. C	45. A	70. C	95. B
21. B	46. C	71. B	96. A
22. B	47. C	72. B	97. D
23. C	48. D	73. A	98. A
24. D	49. C	74. A	99. B
25. A	50. A	75. A and/or B and/or C and/or D	100. C

Explanatory Answers for Practice Examination 1

1. **(B)** The note at the bottom of the picture states that the ground floor is the first floor. Count up from the sidewalk and you will find the fire coming out the fourth-floor windows.

2. **(B)** As you stand in front facing the buildings, the bank is on the left side of the page. The smoke and flames are blowing from the left to the right and up.

3. **(D)** The fire escapes are on the left side of the apartment building and are indicated by the words "fire escape." Count up. The person is on the fire escape at the fifth floor.

4. **(B)** The ground floor is the first floor. Count up again.

5. **(B)** The front window on the fifth floor is directly over the fire and closest to the fire.

6. **(D)** The building on fire is an "isolated" building. This means that there is no direct access to the roof from the roofs of the adjacent buildings.

7. **(A)** As you stand facing the front of the buildings, the fire escapes are on your left-hand side. This is considered the left side of the building for purposes of this exam.

8. **(A)** The fire hydrant is directly in front of the bank.

9. **(D)** The fire escape is in the lower right corner of the diagram. The dining room is in the upper left corner. The dining room is the greatest distance from the fire escape.

10. **(B)** Bedroom 1 has only one door. The living room, dining room, and kitchen each have at least two door openings.

11. **(D)** As indicated in the diagram, the fire escape is directly outside bedroom 3.

12. **(B)** The only door from bathroom 2 leads directly to bedroom 2.

13. **(B)** This is indicated by the door symbol between bathroom 2 and bedroom 2.

14. **(D)** The terrace on the left side of the building has two door symbols to indicate that access from the terrace can be gained to the living room and to the dining room.

15. **(A)** The window symbols are visible in the black outline of the floor plan. This shows that there are two windows in the dining room and three windows in bedroom 2.

16. **(D)** Doorways are shown by the break in the solid lines within the floor plan. The kitchen can be entered using the doorway in the dining room as well as the doorway from the foyer.

17. **(D)** The phrase "should be" is the best phrase to complete the sentence.

18. **(E)** The word "was" is the best word to complete the sentence. Because we are talking about yesterday, the past tense is implied.

19. **(D)** The word "is" is the best word to complete the sentence. Because firefighter Curtis is presently up for promotion, the present tense is used.

20. **(C)** The word "quickly" is the best word to complete the sentence.

21. **(B)** The word "whose" is the best word to complete the sentence. "Whose" refers to firefighter McDaniel.

22. **(B)** As stated in the third sentence of the second paragraph, removal of power from the third rail is the FIRST step firefighters must take when evacuating people through tunnels. By removing the power, the chances of people being electrocuted are removed. This also stops the trains from moving so that people walking on the tracks are not hit by trains.

23. **(C)** This is done to inform the trainmaster of the type of emergency and to ensure that he keeps the power turned off.

24. **(D)** The five white lights in a vertical cluster can easily be seen in the darkened tunnel and can be recognized as an exit from a distance.

25. **(A)** The fire cannot be in a factory because all factories are located between 9th Avenue and 12th Avenue. Private houses are all between one and three stories tall.

26. **(A)** There are both stores and private houses on 47th Street between 6th Avenue and 7th Avenue, but the private houses are at most three stories high. Therefore, the fire must be in a one- or two-story store.

27. **(D)** At alarms 1, 3, and 5 (all fires below the third floor of the building) Firefighter Sims, in the control position, assisted the driver in attaching the hose to the hydrant. At the fires on the third floor or above, he helped in getting the hoseline into the building. The control man also is responsible for making sure that enough hose has been taken to reach the fire.

28. **(C)** The compass directional in the upper left corner of the diagram indicates that travel from 7th Avenue to 6th Avenue is travel in the east-west direction. Corridor D is the east-west hallway in the center of the building. Rooms 107, 109, 125, 126, 127, and 128 are in corridor D.

29. **(C)** The firefighters enter the building on 40th Street, which is the south side of the building. Upon entering the lobby, they would have to turn east in hallway E, north in Hallway B, and west in hallway C.

30. **(D)** From the diagram, we are able to see that Truck 21 is facing north on 7th Avenue. The nearest hose connection is E, about midway between 39th and 40th Streets.

31. **(D)** The description of row frames and brownstones indicates that either can be three stories high and 25 feet wide. The key is in the fire escape clauses. Fire escapes are required on row frame buildings higher than two stories. Rear fire escapes are required in changed brownstones. This cannot be an old law tenement because it is fewer than five stories high.

32. **(B)** As stated in the description of old law tenements, fire escapes are required on the rear of the building if there are two apartments on a floor, and fire escapes are required on both the front and the rear if there are four apartments on a floor. This is done to provide two means of escape from every apartment. Row frames and brownstones do not have front fire escapes.

33. **(C)** If the building is 30 feet wide, it must be a row frame. A row frame of more than two stories must have a rear fire escape.

34. **(C)** This is stated in example 2 of the "MAY DAY" messages. Depending on the extent of the collapse, it is extremely important that an immediate roll call be taken to make sure all firefighters have escaped injury and none are trapped.

35. **(B)** This is explained in example 3 of the "MAY DAY" messages.

36. **(C)** The probable collapse at this point is more serious than the broken wrist. "MAY DAY" should be used instead of "URGENT, URGENT."

37. **(C)** Stairway B is the dark-colored stairway. It enters the fourth floor on the east side. With the fifth floor outlet not usable, this is the nearest usable outlet below the fire.

38. **(C)** By using stairway A on the east side of the seventh floor, he would come out on the eighth floor west.

39. **(D)** If the front entrance of the building faces south and the front of the building is Exposure 1, then in clockwise order, Exposure 2 would face west, Exposure 3 north, and Exposure 4 east.

40. **(D)** The first paragraph states that ventilation enables heated air, smoke, and gases to be replaced with fresh air.

41. **(C)** The fourth paragraph states that property also can be damaged by the water that firefighters use to extinguish a fire.

42. **(E)** The second paragraph states that removing dark smoke and gasses can improve visibility so that firefighters can locate victims faster.

43. **(D)** The third paragraph states that creating a hole in the roof directly above the fire will keep the fire from spreading.

44. **(B)** The advantages to ventilation are listed at the end of the first paragraph. They are advantages in rescue operations, fire extinguishing, and property conservation.

45. **(A)** Hydrants 3 and 4 are both considerably closer to the fire than hydrants 1 and 7. Hydrant 4 cannot be reached without passing through smoke and fire. Hydrant 3 is in the best position for protecting the warehouse from the fire.

46. **(C)** Try each route and see for yourself. This route also gets them out of the complex and away from danger.

47. **(C)** This is indicated on the diagram for the question.

48. **(D)** Of the choices offered, only hydrants 6 and 7 are outside the fence. Hydrant 7 is closer to the fire.

49. **(C)** Hi-Lo Lane is engulfed in smoke and flame. The route in choice (C) is shorter than the route in choice (D).

50. **(A)** Try each route. This is the shortest way.

51. **(B)** If the door is unlocked, it is the simplest and preferred way to enter an apartment. Sometimes, in the anxious moments of arrival at a fire, the firefighter presumes the door to be locked and immediately starts forcible entry. This is a mistake in procedure.

52. **(C)** Because the aerial ladder is in use, the firefighter must go by step 1b.

53. **(A)** This is proper procedure as stated in step 3b and step 4. Opening the skylight will immediately ventilate the interior stairs, causing pent-up smoke and gasses to escape. This makes it easier for firefighting interior forces to operate.

54. **(B)** This is proper procedure as stated in step 2. By doing this, you generally will relieve any fears the people inside may have and might even secure their assistance from their position inside.

55. **(C)** In general, a broken arm is not a life-threatening injury. Rather than trying to remove the person by any means possible, firefighters should adhere to the instruction in step 5 and send for an elevator mechanic who is better equipped to extricate the victim. The firefighters should, however, maintain verbal contact with the operator to make sure he does not go into shock. If the injury situation appears to worsen, they must move directly to step 6b and take measures before the elevator mechanic arrives.

56. **(A)** This is the second choice. Victims to be rescued generally are in a very nervous state. Use of the fire escape would be more familiar to them and would be less stressful than climbing over the railing and down a portable ladder.

57. **(B)** Room 304 is on the south side of the building. Room 306 is the nearest room to the west of Room 304.

58. **(B)** Stairway B, as shown in the diagram, is the closest. The stairway used by the firefighter to gain access to the third floor was stairway A.

59. **(A)** Room 303 is directly across the hallway from Room 304. Room 403 is directly above Room 303.

60. **(A)** By checking the compass points in the center of the diagram, we know that Rooms 303 and 305 are on the south side of the courtyard. Rooms 323 and 325 are on the opposite side of the courtyard, which is the north side of the building.

61. **(D)** Stairway D is the correct answer. The northeast stair is stairway C. By coming up stairway D, both hoselines can operate from the same direction to extinguish the fire.

62. **(A)** Diagram A is correct because the rotating base is directly in front of the victim at the window. Ladders generally are positioned with both beams on the sill to prevent twisting of the ladder under the weight of people.

63. **(B)** This is stated in rule 2. Doors should be locked on the stairway side to force people to exit at the ground level in the case of a fire. In buildings over 100 feet in height, people can seek refuge several floors below the fire floor. Doors on the street level are locked from the street side for security purposes.

64. **(A)** The correct answer is stated in rules 3 and 4. The re-entry signs are posted to safely evacuate occupants to floors below the fire without having occupants walk down many flights of stairs in very tall buildings.

65. **(D)** The answer is directly stated in the third sentence. First removing the victim to street level would only delay the administration of oxygen.

66. **(D)** The question asks about transmittal of alarms, not about receipt. The fourth and fifth sentences give the answer.

67. **(B)** See the first sentence of the first paragraph. The firefighter on housewatch has other duties such as greeting visitors, answering the telephone, and maintaining security, but his or her primary responsibility is to properly receive alarm information.

68. **(C)** Read carefully. The answer is in the next to last sentence.

69. **(C)** See the third sentence.

70. **(C)** Suspect #3 only differs from the witness's description by 10 lbs. in weight. The other suspects all differ on two or more of the descriptive items.

71. **(B)** Suspect #2 differs from the witness's description by 20 lbs. and also has blond rather than brown hair.

72. **(B)** The aerial ladder is at the front of the house, and the portable ladder is at the left. The second company must now ladder right and rear. Ladder company personnel are assigned specific areas to search so that no area of the building is left unsearched.

73. **(A)** Victim 3 is removed first because he is the most seriously exposed to the fire. Victim 2 is easier to remove because she is not disabled. Victim 1 is the last one removed because this removal will be more difficult and will take more time due to the disability.

74. **(A)** A pancake collapse leaves little or no space beneath that is capable of providing refuge for trapped victims.

75. After realizing that the wording of this question might lead to different interpretations of its meaning, the Department of Personnel agreed to accept any answer. With the reading originally intended, the answer should be (A). The definition of a floor collapse requires that there be large pieces of machinery on the floor below that hold up sections of floor and create spaces that provide shelter. Large pieces of machinery on any or every floor are most likely to be found in a factory building. In other buildings, large pieces of machinery are more likely to be found only in the basement. This limits the number of floor collapses possible.

76. **(A)** The third rail carries the electricity that provides power. The contact shoes and metal plates are one and the same. The shoes or plates carry electricity to the motor.

77. **(A)** The answer is in the third sentence.

78. **(D)** See the second paragraph. The reason is that, if the hoseline is taken from the outlet on the fire floor and the fire gains control of the floor and hall, there is the possibility that the line will have to be abandoned. Using the outlet below the fire enables firefighters to work in a smoke-free atmosphere. Should the fire get out into the hallway, they still can maintain control of the water supply.

79. **(B)** Search and rescue, especially with knowledge of trapped victims as in this case, must be the first order of business.

80. **(D)** See the last paragraph. The tool attached to the rope is likely to be an ax, but this is not specified in the passage. The tool attached to the rope gives the firefighter greater reach when breaking windows from above.

81. **(C)** Almost by definition, the roof person would effect a rescue from above the fire. Fourteen-story buildings do not have fire escapes; therefore, rescue from above, except on the highest floors where it might be made from the roof, must be made from an apartment window above the fire.

82. **(C)** The answer is stated in step 3. Checking the permits first makes you aware of:

    1. Whether the permits are up to date.
    2. The areas in the building that require special attention such as fuel oil storage, refrigeration equipment, and paint storage.

    The permit will note information such as the amount to be stored, the number and type of refrigerants being used, and the grade and amount of fuel oil being used and stored.

83. **(A)** The lower limit for propane is listed in the table as 2.2%. Therefore, any percentage less than 2.2% will not burn.

84. **(E)** The upper limit for hydrogen is listed in the table as 75%. Therefore, when the percentage of hydrogen to oxygen is more than 75%, the mixture will not burn.

85. **(C)** In order for natural gas to burn, the percentage must be more than 5.0% and less than 15%. 10% is the only percentage listed that is between these two numbers.

86. **(D)** See step 2a.

87. This question was deleted following candidates' appeal.

88. **(B)** The word that is spelled incorrectly is "ventelation." The correct spelling is "ventilation."

89. **(A)** The word that is spelled incorrectly is "suceptiblility." The correct spelling is "susceptibility."

90. **(B)** If you divide 1 by 5 you get .20.

91. **(C)** If all three section are fully extended, the ladder will reach $18 \times 3 = 54$ feet. However, the ladders are on top of the truck, which is 8 feet tall, so $54 + 8 = 62$ feet.

92. **(C)** For every mile per hour that a fire engine is traveling, it takes $3/4$ of a foot to come to a complete stop. If an engine is traveling 60 mph, the answer is found by multiplying 60 by $3/4$. In other words, $60 \times 3 = 180$; $180/4 = 45$.

93. **(C)** It takes the water 1 second to travel through 5 feet of hose. Therefore, $125/5 = 25$.

94. **(B)** There are 60 minutes in every hour. 5:30 is 1 hour and 10 minutes after 4:20. Therefore, 60 minutes + 10 minutes = 70 minutes.

95. **(B)** If the total number of fires in 1991 was 312, take $312 - (48 + 167)$. First solve what is in the parentheses. You now have $312 - 215 = 97$.

96. **(A)** Car accidents increased from 154 to 186 from 1990 to 1993. First find the difference between these numbers, $186 - 154 = 32$. Because we want to know the percent increase, we want to know what percentage 32 is of 154. As in problem 64, we take $32/154 = .21$. To convert to percent, take $.21 \times 100\% = 21\%$.

97. **(D)** To find the average, 48 + 59 + 43 = 150. 150/3 = 50.

98. **(A)** A firefighter assigned to the roof position who sees the roof sagging and in danger of collapse should immediately give a "MAY DAY" signal. This alerts all firefighters operating on the top floor to evacuate the collapse area at once.

99. **(B)** A person bleeding heavily from a stomach wound is in serious condition. If the bleeding is not halted, the victim can go into shock and can die from loss of blood.

100. **(C)** At this fire, the occupant of the fourth floor should be rescued first. The floor directly above the fire is the most seriously exposed floor. Dangerous products of combustion, mainly carbon monoxide, rise rapidly through convection to fill the area, excluding oxygen and causing death or serious injury to trapped people.

Answer Sheet for Practice Examination 2

1. Ⓐ Ⓑ Ⓒ Ⓓ Ⓔ
2. Ⓐ Ⓑ Ⓒ Ⓓ Ⓔ
3. Ⓐ Ⓑ Ⓒ Ⓓ Ⓔ
4. Ⓐ Ⓑ Ⓒ Ⓓ Ⓔ
5. Ⓐ Ⓑ Ⓒ Ⓓ Ⓔ
6. Ⓐ Ⓑ Ⓒ Ⓓ Ⓔ
7. Ⓐ Ⓑ Ⓒ Ⓓ Ⓔ
8. Ⓐ Ⓑ Ⓒ Ⓓ Ⓔ
9. Ⓐ Ⓑ Ⓒ Ⓓ Ⓔ
10. Ⓐ Ⓑ Ⓒ Ⓓ Ⓔ
11. Ⓐ Ⓑ Ⓒ Ⓓ Ⓔ
12. Ⓐ Ⓑ Ⓒ Ⓓ Ⓔ
13. Ⓐ Ⓑ Ⓒ Ⓓ Ⓔ
14. Ⓐ Ⓑ Ⓒ Ⓓ Ⓔ
15. Ⓐ Ⓑ Ⓒ Ⓓ Ⓔ
16. Ⓐ Ⓑ Ⓒ Ⓓ Ⓔ
17. Ⓐ Ⓑ Ⓒ Ⓓ Ⓔ
18. Ⓐ Ⓑ Ⓒ Ⓓ Ⓔ
19. Ⓐ Ⓑ Ⓒ Ⓓ Ⓔ
20. Ⓐ Ⓑ Ⓒ Ⓓ Ⓔ
21. Ⓐ Ⓑ Ⓒ Ⓓ Ⓔ
22. Ⓐ Ⓑ Ⓒ Ⓓ Ⓔ
23. Ⓐ Ⓑ Ⓒ Ⓓ Ⓔ
24. Ⓐ Ⓑ Ⓒ Ⓓ Ⓔ
25. Ⓐ Ⓑ Ⓒ Ⓓ Ⓔ
26. Ⓐ Ⓑ Ⓒ Ⓓ Ⓔ
27. Ⓐ Ⓑ Ⓒ Ⓓ Ⓔ
28. Ⓐ Ⓑ Ⓒ Ⓓ Ⓔ
29. Ⓐ Ⓑ Ⓒ Ⓓ Ⓔ
30. Ⓐ Ⓑ Ⓒ Ⓓ Ⓔ
31. Ⓐ Ⓑ Ⓒ Ⓓ Ⓔ
32. Ⓐ Ⓑ Ⓒ Ⓓ Ⓔ
33. Ⓐ Ⓑ Ⓒ Ⓓ Ⓔ
34. Ⓐ Ⓑ Ⓒ Ⓓ Ⓔ
35. Ⓐ Ⓑ Ⓒ Ⓓ Ⓔ
36. Ⓐ Ⓑ Ⓒ Ⓓ Ⓔ
37. Ⓐ Ⓑ Ⓒ Ⓓ Ⓔ
38. Ⓐ Ⓑ Ⓒ Ⓓ Ⓔ
39. Ⓐ Ⓑ Ⓒ Ⓓ Ⓔ
40. Ⓐ Ⓑ Ⓒ Ⓓ Ⓔ
41. Ⓐ Ⓑ Ⓒ Ⓓ Ⓔ
42. Ⓐ Ⓑ Ⓒ Ⓓ Ⓔ
43. Ⓐ Ⓑ Ⓒ Ⓓ Ⓔ
44. Ⓐ Ⓑ Ⓒ Ⓓ Ⓔ
45. Ⓐ Ⓑ Ⓒ Ⓓ Ⓔ
46. Ⓐ Ⓑ Ⓒ Ⓓ Ⓔ
47. Ⓐ Ⓑ Ⓒ Ⓓ Ⓔ
48. Ⓐ Ⓑ Ⓒ Ⓓ Ⓔ
49. Ⓐ Ⓑ Ⓒ Ⓓ Ⓔ
50. Ⓐ Ⓑ Ⓒ Ⓓ Ⓔ
51. Ⓐ Ⓑ Ⓒ Ⓓ Ⓔ
52. Ⓐ Ⓑ Ⓒ Ⓓ Ⓔ
53. Ⓐ Ⓑ Ⓒ Ⓓ Ⓔ
54. Ⓐ Ⓑ Ⓒ Ⓓ Ⓔ
55. Ⓐ Ⓑ Ⓒ Ⓓ Ⓔ
56. Ⓐ Ⓑ Ⓒ Ⓓ Ⓔ
57. Ⓐ Ⓑ Ⓒ Ⓓ Ⓔ
58. Ⓐ Ⓑ Ⓒ Ⓓ Ⓔ
59. Ⓐ Ⓑ Ⓒ Ⓓ Ⓔ
60. Ⓐ Ⓑ Ⓒ Ⓓ Ⓔ
61. Ⓐ Ⓑ Ⓒ Ⓓ Ⓔ
62. Ⓐ Ⓑ Ⓒ Ⓓ Ⓔ
63. Ⓐ Ⓑ Ⓒ Ⓓ Ⓔ
64. Ⓐ Ⓑ Ⓒ Ⓓ Ⓔ
65. Ⓐ Ⓑ Ⓒ Ⓓ Ⓔ
66. Ⓐ Ⓑ Ⓒ Ⓓ Ⓔ
67. Ⓐ Ⓑ Ⓒ Ⓓ Ⓔ
68. Ⓐ Ⓑ Ⓒ Ⓓ Ⓔ
69. Ⓐ Ⓑ Ⓒ Ⓓ Ⓔ
70. Ⓐ Ⓑ Ⓒ Ⓓ Ⓔ
71. Ⓐ Ⓑ Ⓒ Ⓓ Ⓔ
72. Ⓐ Ⓑ Ⓒ Ⓓ Ⓔ
73. Ⓐ Ⓑ Ⓒ Ⓓ Ⓔ
74. Ⓐ Ⓑ Ⓒ Ⓓ Ⓔ
75. Ⓐ Ⓑ Ⓒ Ⓓ Ⓔ
76. Ⓐ Ⓑ Ⓒ Ⓓ Ⓔ
77. Ⓐ Ⓑ Ⓒ Ⓓ Ⓔ
78. Ⓐ Ⓑ Ⓒ Ⓓ Ⓔ
79. Ⓐ Ⓑ Ⓒ Ⓓ Ⓔ
80. Ⓐ Ⓑ Ⓒ Ⓓ Ⓔ
81. Ⓐ Ⓑ Ⓒ Ⓓ Ⓔ
82. Ⓐ Ⓑ Ⓒ Ⓓ Ⓔ
83. Ⓐ Ⓑ Ⓒ Ⓓ Ⓔ
84. Ⓐ Ⓑ Ⓒ Ⓓ Ⓔ
85. Ⓐ Ⓑ Ⓒ Ⓓ Ⓔ
86. Ⓐ Ⓑ Ⓒ Ⓓ Ⓔ
87. Ⓐ Ⓑ Ⓒ Ⓓ Ⓔ
88. Ⓐ Ⓑ Ⓒ Ⓓ Ⓔ
89. Ⓐ Ⓑ Ⓒ Ⓓ Ⓔ
90. Ⓐ Ⓑ Ⓒ Ⓓ Ⓔ
91. Ⓐ Ⓑ Ⓒ Ⓓ Ⓔ
92. Ⓐ Ⓑ Ⓒ Ⓓ Ⓔ
93. Ⓐ Ⓑ Ⓒ Ⓓ Ⓔ
94. Ⓐ Ⓑ Ⓒ Ⓓ Ⓔ
95. Ⓐ Ⓑ Ⓒ Ⓓ Ⓔ
96. Ⓐ Ⓑ Ⓒ Ⓓ Ⓔ
97. Ⓐ Ⓑ Ⓒ Ⓓ Ⓔ
98. Ⓐ Ⓑ Ⓒ Ⓓ Ⓔ
99. Ⓐ Ⓑ Ⓒ Ⓓ Ⓔ
100. Ⓐ Ⓑ Ⓒ Ⓓ Ⓔ

TEAR HERE

PRACTICE EXAMINATION 2

The time allowed for the entire examination is 3½ hours.

DIRECTIONS: *Each question has four suggested answers lettered A, B, C, and D. Decide which one is the best answer. On the sample answer sheet, find the question number that corresponds to the answer you have selected and darken the area with a soft pencil.*

Memory Question

Firefighters must possess the ability to readily recall and identify the location of fire protection appliances, unusual building occupancies, building construction features, and other important physical attributes present at the site of a building fire. This knowledge is important for the safety of firefighters, and it enables them to be more efficient.

You will have five minutes to study and memorize the details of the buildings on Pine Street. Questions 1 through 13, beginning below, are based on the picture and the legend. You may not refer back to either the picture or the legend while answering these questions.

DIRECTIONS: *Answer questions 1 through 13 based on the picture and legend of Pine Street found on page 212. Remember that you may NOT refer back to either the picture or the legend while answering the questions.*

1. The building that would require the longest ladder to enable you to climb to the top floor would most likely be
   (A) 212 Pine Street
   (B) 216 Pine Street
   (C) 200 Pine Street
   (D) 220 Pine Street

2. If a fire occurred in a building on Pine Street at 3 A.M., the greatest life hazard would most likely be in the building at
   (A) 212 Pine Street
   (B) 218 Pine Street
   (C) 220 Pine Street
   (D) 216 Pine Street

3. For a building fire on Pine Street, if the fire department pumper is connected to the hydrant, the firefighter would need the longest stretch of hose to reach the
   (A) Ace Dress Co.
   (B) Pine Diner
   (C) Auto Supplies Store
   (D) Steel Door Co.

4. A firefighter is opening the skylights to remove smoke and heat from the building. You would be correct to assume that he is on the roof of the
   (A) lumber company
   (B) dress company
   (C) hardware store
   (D) auto supply store

5. As a firefighter, you are ordered by Lieutenant Brown to connect a hoseline to the automatic sprinkler siamese connection. You would find this connection in front of
   (A) 220 Pine Street
   (B) 216 Pine Street
   (C) 200 Pine Street
   (D) 210 Pine Street

6. Roll-down metal shutters are installed on storefronts to deter burglars. They might cause a fire to burn undetected, however, until it has gained considerable headway. The tenant on this block with roll-down metal shutters is the
   (A) hardware store
   (B) diner
   (C) lumber company
   (D) auto supply store

7. A firefighter is cutting a hole in the roof close to the front wall of 218 Pine Street. Severe conditions on the roof require that the firefighters make a fast exit from the roof. The safest and quickest exit can be made by
   (A) walking across the roof to 216 Pine Street
   (B) stepping over the parapet to the roof of 220 Pine Street
   (C) jumping across to the building behind 218 Pine Street
   (D) waiting until a ladder is placed against the building

8. The best protection against fire in its beginning stage is the automatic wet sprinkler system. The tenant on Pine Street that has the automatic wet sprinkler system is the
   (A) dress company
   (B) hardware store
   (C) lumber company
   (D) Dexter Hotel

9. When extinguishing a smoky fire, it often is advantageous to operate from the windward side of the fire (with the wind at your back). From the diagram of Pine Street, you have observed that the wind is blowing from the
   (A) north
   (B) south
   (C) east
   (D) west

10. Hose stream penetration is important when extinguishing fires. The building on Pine Street that would require the deepest penetration of a hose stream is occupied by
    (A) the Steel Door Co.
    (B) Brown Lumber
    (C) Pine Diner
    (D) the auto supply store

11. Generally speaking, a curved roof is a trussed roof. Trussed roofs are notorious for rapid failure when subjected to high heat and fire for a short period of time. The building on Pine Street with a trussed roof is the
    (A) hardware store
    (B) lumber company
    (C) Steel Door Co.
    (D) Ace Dress Co.

12. Baled material and fibers become extremely heavy when wet. This causes overloading of floors that can result in collapse of the building. Effective use of water would be necessary for a fire at
    (A) 200 Pine Street
    (B) 212 Pine Street
    (C) 214 Pine Street
    (D) 218 Pine Street

13. Protecting exposures is an important part of firefighting strategy. If a heavy fire condition exists in Brown Lumber and your strategy is to protect the adjoining building, you most likely would have your firefighters stretch a hoseline into the
    (A) Dexter Hotel
    (B) Ace Dress Co.
    (C) Pine Diner
    (D) hardware store

14. A problem likely to be found in a hotel fire that is not usually found in fires involving other types of residences, such as apartment houses, is
    (A) obstructions in hallways and other passageways
    (B) large numbers of persons in relation to the number of rooms
    (C) delay in the transmission of the fire alarm
    (D) many occupants who don't know the location of exits

15. On the way to work one morning, a firefighter notices a high-tension wire that has blown down and is lying across the sidewalk and into the road. The most appropriate course of action for the firefighter to take would be to
    (A) move the wire to one side by means of a stock or branch of a tree and continue on his way
    (B) continue on his way to the firehouse and report the situation to the officer on duty
    (C) call the public utility company from the first public telephone he passes on his way to the firehouse
    (D) stand by the wire to warn away passersby and ask one of them to call the public utility company

16. Generally, firefighters on prevention inspection duty do not inspect the living quarters of private dwellings unless the occupants agree to the inspection. The best explanation why private dwellings are excluded from compulsory inspections is that
    (A) private dwellings seldom catch fire
    (B) fires in private dwellings are more easily extinguished than other types of fires
    (C) people might resent such inspections as an invasion of privacy
    (D) the monetary value of private dwellings is lower than that of other types of occupancies

17. Fire lines usually are established by the police to keep bystanders out of the immediate vicinity while firefighters are fighting the fire. Of the following, the best justification for the establishment of these fire lines is to
    (A) prevent theft of property from partially destroyed apartments or stores
    (B) prevent interference with firefighter operations
    (C) give privacy to the victims of the fire
    (D) help apprehend arsonists in the crowd

18. What is 12 squared?
    (A) 144   (B) 169   (C) 196   (D) 225   (E) 256

19. At a large fire outside the city, three water tankers have arrived to supply water to the firefighters. Tanker A holds 1,500 gallons, Tanker B holds 2,500 gallons, and Tanker C holds 3,200 gallons of water. How many gallons of water are available to the firefighters?
    (A) 4,000 gallons
    (B) 4,700 gallons
    (C) 5,700 gallons
    (D) 6,400 gallons
    (E) 7,200 gallons

20. A fire engine is required to carry six hoses. The lengths of the hoses are 100 feet, 100 feet, 150 feet, 150 feet, 250 feet, and 300 feet. What is the average hose length?
    (A) 300 feet   (B) 275 feet   (C) 200 feet   (D) 175 feet   (E) 150 feet

21. A firefighter needs to crawl through a window that is 5 feet high and 2 feet wide. What is the area of the opening?
    (A) 30 square feet
    (B) 25 square feet
    (C) 15 square feet
    (D) 10 square feet
    (E) 5 square feet

DIRECTIONS: *Use the following table to answer questions 22 and 23.*

Four different categories of fire engine carry ladders or platforms. Each of these has a specific range when the ladder or platform is extended.

Category	Range
Aerial ladder	50 to 135 feet
Aerial ladder platform	85 to 110 feet
Telescoping aerial platform	50 to 100 feet
Articulating aerial platform	55 to 102 feet

22. Which of the four categories has the largest range?
    (A) Aerial ladder platform
    (B) Aerial ladder
    (C) Articulating ladder platform
    (D) Telescoping aerial platform
    (E) Articulating aerial platform

23. If the articulating aerial platform were extended to 75 feet, what percentage of its range would be in use?
    (A) 42.5%
    (B) 43.5%
    (C) 45.3%
    (D) 46.2%
    (E) 47.6%

24. Visitors near patients in oxygen tents are not permitted to smoke. The best of the following reasons for this prohibition is that
    (A) the flame of the cigarette or cigar might flare dangerously
    (B) smoking tobacco is irritating to persons with respiratory diseases
    (C) smoking in bed is one of the major causes of fire
    (D) diseases can be transmitted by means of tobacco smoke

25. At a hot and smoky fire, a lieutenant orders firefighters to work in pairs. Of the following, the best justification for this order is that
    (A) better communications result because one person can bring messages back to the lieutenant
    (B) more efficient operation results because many vital activities require two people
    (C) better morale results because firefighters are more willing to face danger in pairs
    (D) safer operations result because one firefighter can help if the other becomes disabled

26. Firefighters frequently open windows, doors, and skylights of a building on fire in a planned or systematic way to ventilate the building. The least likely to be accomplished by ventilation is the
    (A) increase in visibility of firefighters
    (B) slowdown in the rate of burning of the building
    (C) reduction in the danger from toxic gases
    (D) control of the direction of travel of the fire

27. At the first sign of a fire, the manager of a movie theater had the lights turned on and made the following announcement: "Ladies and gentlemen, the management has found it necessary to dismiss the audience. Please remain seated until it is time for your aisle to file out. In leaving the theater, follow the directions of the ushers. There is no danger involved." The manager's action in this situation was
    (A) proper
    (B) improper, chiefly because the manager did not tell the audience the reason for the dismissal
    (C) improper, chiefly because the manager did not permit all members of the audience to leave at once
    (D) improper, chiefly because the manager misled the audience by saying that there was no danger

28. Generally, sprinkler heads must be replaced each time they are used. The best explanation for this is that the sprinkler heads
    (A) are subject to rusting after discharging water
    (B) might become clogged after discharging water
    (C) have a distorted pattern of water discharge after use
    (D) are set off by the effect of heat on metal and cannot be reset

29. After a fire in an apartment had been brought under control, firefighters were engaged in extinguishing the last traces of fire. A firefighter who noticed an expensive vase in the room in which activities were concentrated moved it to an empty closet in another room. The firefighter's action was
    (A) proper, chiefly because the owners would realize that extreme care was taken to avoid damage to their possessions
    (B) improper, chiefly because disturbance of personal possessions should be kept to a minimum
    (C) proper, chiefly because the chance of damage is reduced
    (D) improper, chiefly because the firefighter should have devoted this time to putting out the fire

30. While standing in front of a firehouse, a firefighter is approached by a woman with a baby carriage. The woman asks the firefighter if he will "keep an eye" on the baby while she visits a doctor in a nearby building. In this situation, the best course of action for the firefighter to take is to
    (A) agree to the woman's request but warn her that it might be necessary to leave to answer an alarm
    (B) refuse politely after explaining that he is on duty and cannot become involved in other activities
    (C) refer the woman to the officer on duty
    (D) ask the officer on duty for permission to grant the favor

31. Suppose that, while cooking, a pan of grease catches fire. The method that would be most effective, if available, in putting out the fire is to
    (A) dash a bucket of water on the fire
    (B) direct a stream of water on the fire from a fire extinguisher
    (C) pour a bottle of household ammonia water on the fire
    (D) empty a box of baking soda on the fire

32. Persons engaged in certain hazardous activities are required to obtain a fire department permit or certificate for which a fee is charged. The main reason for requiring permits or certificates is to
    (A) obtain revenue for the city government
    (B) prevent unqualified persons from engaging in these activities
    (C) obtain information about these activities to plan for fire emergencies
    (D) warn the public of the hazardous nature of these activities

33. A firefighter traveling to work is stopped by a citizen who complains that the employees of a nearby store frequently pile empty crates and boxes in a doorway, blocking passage. The most appropriate action for the firefighter to take is to
    (A) assure the citizen that the fire department's inspection activities will eventually catch up with the store
    (B) obtain the address of the store and investigate to determine whether the citizen's complaint is justified
    (C) obtain the address of the store and report the complaint to a superior officer
    (D) ask the citizen for specific dates on which this practice occurred to determine whether the complaint is justified

34. The crime of arson is defined as the willful burning of a house or other property. The best illustration of arson is a fire
    (A) in an apartment started by a four-year-old boy playing with matches
    (B) in a barn started by a drunken man who overturns a lantern
    (C) in a store started by the bankrupt owner to collect insurance
    (D) in a house started by a neighbor who carelessly burned leaves in the garden

35. The fire department now uses companies on fire duty with their apparatus for fire prevention inspection in commercial buildings. The change most important in making this inspection procedure practicable was the
    (A) reduction of hours of work for firefighters
    (B) use of two-way radio equipment
    (C) use of enclosed cabs on fire apparatus
    (D) increase in property value in the city

36. Many fires are caused by improper use of oxyacetylene torches. The main cause of such fires is the
    (A) high pressure under which the gases are stored
    (B) failure to control or extinguish sparks
    (C) high temperatures generated by the equipment
    (D) explosive nature of the gases used

37. The most important reason for having members of the fire department wear uniforms is to
    (A) indicate the semimilitary nature of the fire department
    (B) build morale and *esprit de corps* of members
    (C) identify members on duty to the public and other members
    (D) provide clothing suitable for the work performed

38. Of the following types of fires, the one that is likely to have the *least* amount of damage from water used in extinguishing it is a fire in
  (A) a rubber toy factory
  (B) a retail hardware store
  (C) an outdoor lumberyard
  (D) a furniture warehouse

39. "In a case of suspected arson, it is important for firefighters engaged in fighting the fire to remember conditions that existed at the time of their arrival. Particular attention should be given to all doors and windows." The main justification for this statement is that knowledge of the condition of the doors and windows might indicate
  (A) where the fire started
  (B) who set the fire
  (C) the best way to ventilate the building
  (D) that someone wanted to prevent it from being extinguished

40. "After a fire has been extinguished, every effort should be made to determine how the fire started." Of the following, the chief reason for determining the origin of the fire is to
  (A) reduce the amount of damage caused by the fire
  (B) determine how the fire should have been fought
  (C) eliminate causes of fire in the future
  (D) explain delays in fighting the fire

41. A firefighter inspecting buildings in a commercial area came to one building whose outside surface appeared to be of natural stone. The owner told the firefighter that it was not necessary to inspect the building because it was "fireproof." The firefighter, however, completed inspection of the building. Of the following, the best reason for continuing the inspection is that
  (A) stone buildings catch fire as readily as wooden buildings
  (B) the fire department cannot make exceptions in its inspection procedures
  (C) the building might have been built of imitation stone
  (D) interiors and contents of stone buildings can catch fire

42. The *least* valid reason for the fire department to investigate the causes of fire is to
  (A) determine whether the fire was the result of arson
  (B) estimate the amount of loss for insurance purposes
  (C) gather information useful in fire prevention
  (D) discover violations of the Fire Prevention Code

43. While on duty at a fire, a probationary firefighter receives an order from the lieutenant that appears to conflict with the principles of firefighting taught at the fire school. Of the following, the best course of action for the firefighter to take is to follow the order and, at a convenient time after the fire, to
  (A) discuss the apparent inconsistency with the lieutenant
  (B) discuss the apparent inconsistency with another officer
  (C) mention this apparent inconsistency in an informal discussion
  (D) ask a more experienced firefighter about the apparent inconsistency

44. When fighting fires on piers, the fire department frequently drafts salt water from the harbor. The chief advantage of using harbor water instead of relying on water from street mains is that harbor water is
    (A) less likely to cause water damage
    (B) available in unlimited quantities
    (C) more effective in extinguishing fires due to its salt content
    (D) less likely to freeze in low temperatures due to its salt content

DIRECTIONS: *Questions 45 through 48 are based on the following paragraph.*

The canister type of gas mask consists of a tight-fitting face piece connected to a canister containing chemicals that filter toxic gases and smoke from otherwise breathable air. These masks are of value when used with due regard to the fact that two or three percent of gas in air is about the highest concentration that the chemicals in the canister will absorb and that these masks do not provide the oxygen necessary for the support of life. In general, if flame is visible, there is sufficient oxygen for firefighters although toxic gases might be present. Where there is heavy smoke and no flame, an oxygen deficiency might exist. Fatalities have occurred where filter-type canister masks have been used in attempting rescue from manholes, wells, basements, or other locations deficient in oxygen.

45. If the mask described in the paragraph is used in an atmosphere containing oxygen, nitrogen, and carbon monoxide, we would expect the mask to remove from the air breathed
    (A) the nitrogen only
    (B) the carbon monoxide only
    (C) the nitrogen and the carbon monoxide
    (D) none of these gases

46. According to this paragraph, when firefighters are wearing these masks at a fire in which flame is visible, the firefighters generally can feel that, as far as breathing is concerned, they are
    (A) safe, because the mask will provide them with sufficient oxygen to live
    (B) unsafe, unless the gas concentration is below two or three percent
    (C) safe, provided the gas concentration is above two or three percent
    (D) unsafe, because the mask will not provide them with sufficient oxygen to live

47. According to this paragraph, fatalities have occurred to persons using this type of gas mask in manholes, wells, and basements because
    (A) the supply of oxygen provided by the mask ran out
    (B) the air in those places did not contain enough oxygen to support life
    (C) heavy smoke interfered with the operation of the mask
    (D) the chemicals in the canister did not function properly

48. The following formula can be used to show, in general, the operation of the gas mask described in the preceding paragraph:

(Chemicals in canister) → (air + gases) = breathable air

The arrow in the formula, when expressed in words, means most nearly

(A) replace
(B) are changed into
(C) act upon
(D) give off

DIRECTIONS: Questions 49 through 51 are based on the following paragraph.

The only openings permitted in fire partitions, except openings for ventilating ducts, shall be those required for doors. There shall be but one such door opening unless the provision of additional openings would not exceed in total width of all doorways 25 percent of the length of the wall. The minimum distance between openings shall be three feet. The maximum area for such a door opening shall be 80 square feet, except that such openings for the passage of trucks may be a maximum of 140 square feet.

49. According to the paragraph, openings in fire partitions are permitted only for
(A) doors
(B) doors and windows
(C) doors and ventilation ducts
(D) doors, windows, and ventilation ducts

50. In a fire partition 22 feet long and 10 feet high, the maximum number of doors 3 feet wide and 7 feet high is
(A) 1
(B) 2
(C) 3
(D) 4

51. The most accurate statement about the preceding layout is that the
(A) total width of the openings is too large
(B) truck opening is too large
(C) truck and door openings are too close together
(D) layout is acceptable

52. Of the following, the basic purpose of fire department inspection of private property is to
(A) make sure that fire department equipment is properly maintained
(B) secure proper maintenance of all entrance and exit facilities
(C) obtain correction of conditions creating undue fire hazards
(D) make sure that walls are properly plastered and painted to prevent the spread of fire

53. Of the following items commonly found in a household, the one that uses the most electric current is
    (A) a 150-watt light bulb
    (B) a toaster
    (C) a door buzzer
    (D) an 8-inch electric fan

DIRECTIONS: *Questions 54 through 56 are based on the following paragraph.*

The average daily flow of water through public water systems in American cities ranges generally between 40 and 250 gallons per capita, depending on the underground leakage in the system, the amount of waste in domestic premises, and the quantity used for industrial purposes. The problem of supplying this water has become serious in many cities. Supplies, which were once adequate, in many cases have become seriously deficient due to greater demands with increased population and growing industrial use of water. Waterworks, operating on fixed schedules of water charges, in many cases have not been able to afford the heavy capital expenditures necessary to provide adequate supply, storage, and distribution facilities. Thus, the adequacy of a public water supply for fire protection in any given location cannot properly be taken for granted.

54. The following four programs are possible ways in which American communities might try to reduce the seriousness of the water shortage problem. The one program that does *not* directly follow from the paragraph is that of
    (A) regular replacement of old street water mains by new ones
    (B) inspection and repair of leaky plumbing fixtures
    (C) fire prevention inspection and education to reduce the amount of water used to extinguish fires
    (D) research into industrial processes to reduce the amount of water used in those processes

55. The main conclusion reached by the author of the paragraph is that
    (A) there is a waste of precious natural resources in America
    (B) communities have failed to control the industrial use of water
    (C) a need exists for increasing the revenue of waterworks to build up adequate supplies of water
    (D) fire departments cannot assume they will always have the necessary supply in water available to fight fires

56. Per capita consumption of water in a community is determined by the formula
    (A) $\dfrac{\text{Population}}{\text{Total consumption in gallons}} = \text{Per capita consumption in gallons}$
    (B) $\dfrac{\text{Total consumption in gallons}}{\text{Population}} = \text{Per capita consumption in gallons}$
    (C) Total consumption in gallons × Population = Per capita consumption in gallons
    (D) Total consumption in gallons − Population = Per capita consumption in gallons

DIRECTIONS: *Use the information in the following passage to answer questions 57 through 60.*

Most accidents that lead to firefighter injuries or fatalities occur at the scene of an emergency. To avoid accidents, it is important to have a system that clarifies the roles and responsibilities of personnel.

The Incident Command System (ICS) is a system that was adopted by the National Fire Academy (NFA) to coordinate resources during a variety of emergency situations. The ICS assists different agencies in coordinating their efforts by using common terminology and operating procedures. The five major functional areas of the ICS are command, operations, planning, logistics, and finance.

Command: For a large-scale operation to run safely and effectively, it is necessary to have people in charge. Individuals occupying positions such as safety officer, liaison, and public information officer are responsible for the ordering and releasing of resources.

Operations: Once command has established the strategic goals, operations is responsible for managing all the operations that directly effect the primary mission. Operations is divided into branches that can be divided again into groups assigned to a particular task such as rescue or ventilation.

Planning: As the incident develops, this area is responsible for the collection, evaluation, dissemination, and use of relevant information. Command will adjust the ordering and releasing of resources according to the information compiled by planning.

Logistics: This area is responsible for supporting the incident through its support and service branches. The support branch maintains supplies, facilities, and vehicle services. The service branch maintains medical, communications, and food services.

Finance: This section takes care of all costs and financial aspects of the incident. Finance is usually only a concern, however, during large-scale incidents that last for a particularly long period of time.

57. The title that best describes the content of this passage is
 (A) "The Purpose and Origin of the National Fire Academy (NFA)"
 (B) "Why Most Injuries and Fatalities Occur at the Scene of an Emergency"
 (C) "The Purpose and Functions of the Incident Command System (ICS)"
 (D) "Why Command is the Most Important Functional Area of the ICS"
 (E) "The Different Branches of Operations"

58. It is necessary to have _____ for a large-scale operation to run safely and effectively.
 (A) people in charge
 (B) adequate ventilation
 (C) sufficient funds
 (D) food services
 (E) firefighter injuries

59. If an operation does not turn out to take much time to resolve, which functional area of the ICS is least likely to be necessary?
    (A) command
    (B) operations
    (C) planning
    (D) logistics
    (E) finance

60. Logistics is NOT in charge of maintaining what service?
    (A) medical
    (B) rescue
    (C) vehicle
    (D) supplies
    (E) communications

DIRECTIONS: Use the information in the following passage to answer questions 61 through 64.

When arson is the suspected cause of a fire, it is important to ensure the proper collection of evidence at the fire scene. The fire investigator typically is responsible for collecting evidence at the scene. It might be the case, however, that an investigator is not available immediately after a fire. In these situations, it is important for the firefighters on the scene to ensure the proper collection of evidence.

The fire department has the authority to control a fire scene for as long after a fire as is necessary. As soon as the last firefighter leaves the scene, however, the department's control of it is limited. Often times, a search warrant or written consent is necessary to make later visits to the scene. It is therefore important that all the evidence be tagged, marked, and photographed as soon as possible. If it is necessary to remove any evidence from the scene, firefighters should be careful not to damage any of the evidence in the process.

Until the investigation of the fire is complete, no one should be allowed to enter the scene unless accompanied by a fire officer. Any entry, even by the owner of the premises, should be carefully logged. The person's name, time of entry, time of departure, and descriptions of any items taken from the premises should be recorded.

If it is deemed necessary, fire scenes can be guarded to ensure the safety of anyone near the scene and to ensure that no one tampers with the evidence. Depending on the number of possible entrances to the scene and the severity of safety concerns, a fire scene could be watched by one person or a full-time guard force.

61. If an investigator is not available immediately after a fire, firefighters should
    (A) destroy the evidence at the scene
    (B) tag, mark, and photograph the evidence
    (C) allow the owner to find the evidence alone
    (D) leave the fire scene unguarded
    (E) firefighters have no control over the evidence

62. All of the following should be logged when someone enters the fire scene except
    (A) a description of the person
    (B) the person's name
    (C) time of entry
    (D) time of departure
    (E) a description of items taken

63. The fire department has the authority to control a fire scene _____.
    (A) for about two weeks
    (B) after proper collection of evidence
    (C) for as long after a fire as necessary
    (D) when the last firefighter leaves
    (E) if the scene is adequately guarded

64. To ensure that the fire scene is not contaminated by the activity of the firefighters, they should NOT
    (A) secure the fire scene
    (B) call the fire investigator
    (C) damage any evidence
    (D) tag any evidence
    (E) be careful with equipment

65. Sand and ashes frequently are placed on icy pavements to prevent skidding. The effect of the sand and ashes is to increase
    (A) inertia
    (B) gravity
    (C) momentum
    (D) friction

66. The air near the ceiling of a room usually is warmer than the air near the floor because
    (A) there is better air circulation at floor level
    (B) warm air is lighter than cold air
    (C) windows usually are nearer the floor than the ceiling
    (D) heating pipes usually run along the ceiling

67. It is safer to use the ladder positioned as shown in diagram 1 than as shown in diagram 2 because in diagram 1

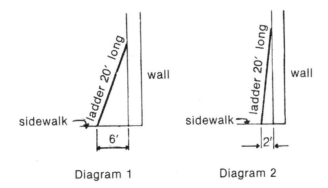

Diagram 1   Diagram 2

(A) less strain is placed on the center rungs of the ladder
(B) it is easier to grip and stand on the ladder
(C) the ladder reaches a lower height
(D) the ladder is less likely to tip over backwards

68. A substance that is subject to spontaneous heating is one that
    (A) is explosive when heated
    (B) is capable of catching fire without an external source of heat
    (C) acts to speed up the burning of material
    (D) liberates oxygen when heated

69. "Chief officers shall instruct their firefighters that, when transmitting particulars of a fire, they shall include the fact that foods are involved if the fire involves premises where foodstuffs are sold or stored." Of the following, the best justification for this regulation is that, when foodstuffs are involved in a fire,
    (A) the fire can reach serious proportions
    (B) police protection might be desirable to prevent looting
    (C) relatively little firefighting equipment may be needed
    (D) inspection to detect contamination may be desirable

70. A fire can continue when fuel, oxygen from the air or another source, and a sufficiently high temperature to maintain combustion are present. The method *most commonly* used to extinguish fires is to
    (A) remove the fuel
    (B) exclude the oxygen from the burning material
    (C) reduce the temperature of the burning material
    (D) smother the flames of the burning material

71. Of the following, the siphon arrangement that would most quickly transfer a solution from the container on the left side to the one on the right side is numbered

    (A) 1
    (B) 2
    (C) 3
    (D) 4

72. Static electricity is a hazard in industry chiefly because it can cause
    (A) dangerous or painful burns
    (B) chemical decomposition of toxic elements
    (C) sparks that can start an explosion
    (D) overheating of electrical equipment

73. A rowboat will float deeper in fresh water than in salt water because
    (A) in the salt water, the salt will occupy part of the space
    (B) fresh water is heavier than salt water
    (C) salt water is heavier than fresh water
    (D) salt water offers less resistance than fresh water

74. In the following diagram, it is easier to get the load onto the platform by using the ramp than by lifting it directly onto the platform. This is true because the effect of the ramp is to

    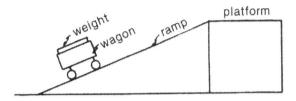

    (A) reduce the amount of friction so that less force is required
    (B) distribute the weight over a larger area
    (C) support part of the load so that less force is needed to move the wagon
    (D) increase the effect of the moving weight

75. More weight can be lifted by the method shown in diagram 2 than the method in diagram 1 because

    (A) it takes less force to turn a crank than it does to pull in a straight line
    (B) the drum will prevent the weight from falling by itself
    (C) the length of the crank is larger than the radius of the drum
    (D) the drum has more rope on it, easing the pull

76. Two balls of the same size but different weights are dropped from a 10-foot height. The most accurate statement is that
    (A) both balls will reach the ground at the same time because they are the same size
    (B) both balls will reach the ground at the same time because the effect of gravity is the same on both balls
    (C) the heavier ball will reach the ground first because it weighs more
    (D) the lighter ball will reach the ground first because air resistance is greater on the heavier ball

77. "A partially filled gasoline drum is a more dangerous fire hazard than a full one." Of the following, the best justification for this statement is that
    (A) a partially filled gasoline drum contains relatively little air
    (B) gasoline is difficult to ignite
    (C) when a gasoline drum is full, the gasoline is more explosive
    (D) gasoline vapors are more explosive than gasoline itself

78. The following diagrams show flywheels made of the same material, with the same dimensions, and attached to similar engines. The solid areas represent equal weights attached to the flywheel. If all three engines are running at the same speed for the same length of time and the power to the engines is shut off simultaneously,

    1   2   3

    (A) wheel 1 will continue turning longest
    (B) wheel 2 will continue turning longest
    (C) wheel 3 will continue turning longest
    (D) all three wheels will continue turning for the same time

79. The substance that expands when freezing is
    (A) alcohol
    (B) ammonia
    (C) mercury
    (D) water

DIRECTIONS: Use the information in the following table to answer questions 80 and 81.

FIRE CLASSIFICATION SYSTEM

Type of Fire	Combustible Materials	Extinguishing Process
Class A	paper, cloth, wood, rubber, plastics	water, foam
Class B	paints, oil, gasoline, mineral spirits	oxygen exclusion (the blanketing effect)
Class C	energized electrical equipment	dry chemicals, carbon dioxide
Class D	combustible metals (aluminum, titanium)	depends on the type of metal

80. It would be dangerous to use water to extinguish a Class C fire because
    (A) it might cause the fire to spread
    (B) there is a risk of electrical shock
    (C) water is only for Class B fires
    (D) the metals are combustible
    (E) water is not to be used for plastics

81. If a painter is working near an open flame and the paint catches fire, the extinguishing process she should use
    (A) depends on the type of paint
    (B) is dry chemicals or carbon dioxide
    (C) is water or a noncombustible foam
    (D) is oxygen exclusion (blanket effect)
    (E) is combustible metals (aluminum)

82. When drilling a hole through a piece of wood with an auger bit, it is considered good practice to clamp a piece of scrap wood to the underside of the piece through which the hole is being drilled. The main reason for this is to
    (A) direct the auger bit
    (B) speed up the drilling operation
    (C) prevent the drill from wobbling
    (D) prevent splintering of the wood

83. "Firefighters holding a life net should keep their eyes on the person jumping from a burning building." Of the following, the best justification for this statement is that
    (A) a person attempting to jump into a life net might overestimate the distance of the net from the building
    (B) some people will not jump into a life net unless given confidence
    (C) a person jumping into a life net might be seriously injured if the net is not allowed to give slightly at the moment of impact
    (D) firefighters holding a life net should be evenly spaced around the net at the moment of impact to distribute the shock

84. If boiling water is poured into a drinking glass, the glass is likely to crack. If a metal spoon is first placed in the glass, however, the glass is much less likely to crack. The reason that the glass with the spoon is less likely to crack is that the spoon
    (A) distributes the water over a larger surface of the glass
    (B) quickly absorbs heat from the water
    (C) reinforces the glass
    (D) reduces the amount of water that can be poured into the glass

85. It takes more energy to force water through a long pipe than through a short pipe of the same diameter. The principal reason for this is
    (A) gravity
    (B) friction
    (C) inertia
    (D) cohesion

86. A pump discharging at 300 pounds per square inch pressure delivers water through 100 feet of pipe laid horizontally. If the valve at the end of the pipe is shut so that no water can flow, the pressure at the valve is, for practical purposes,
    (A) greater than the pressure at the pump
    (B) equal to the pressure at the pump
    (C) less than the pressure at the pump
    (D) greater or less than the pressure at the pump, depending on the type of pump used

87. The leverage system in the following sketch is used to raise a weight. To reduce the amount of force required to raise the weight, it is necessary to

    (A) decrease the length of the lever
    (B) place the weight closer to the fulcrum
    (C) move the weight closer to the person applying the force
    (D) move the fulcrum farther from the weight

DIRECTIONS: *Questions 88 through 91 relate to the information given in the following paragraph and should be answered in accordance with this information.*

Old buildings are liable to possess a special degree of fire risk merely because they are old. Outdated electrical wiring systems and installation of new heating appliances for which the building was not designed can contribute to the increased hazard. Old buildings often have been altered many times, parts of the structure might antedate building codes, and dangerous defects may have been covered up. On the average, old buildings contain more lumber than comparable new buildings. This, in itself, makes old buildings more susceptible to fire. It is not true, though, that sound lumber in old buildings is drier than new lumber. Moisture content of lumber varies with that of the atmosphere to which it is exposed.

88. According to the paragraph, old buildings present a special fire hazard chiefly because of the
    (A) poor planning of the buildings when first designed
    (B) haphazard alteration of the buildings
    (C) dryness of old lumber
    (D) inadequate enforcement of the building codes

89. We can conclude from the paragraph that lumber
    (A) should not be used in buildings unless absolutely necessary
    (B) should not be used near electrical equipment
    (C) is more inflammable than newer types of building materials
    (D) tends to lose its moisture at a constant rate

90. According to the paragraph, the amount of moisture in the wooden parts of a building depends on the
  (A) age of the building
  (B) moisture in the surrounding air
  (C) type of heating equipment used in the building
  (D) quality of lumber used

91. In regard to building codes, the paragraph implies that
  (A) old buildings are exempt from the provisions of building codes
  (B) some buildings now in use were built before building codes were adopted
  (C) building codes usually don't cover electrical wiring systems
  (D) building codes generally are inadequate

92. A firefighter must become familiar with first-aid methods. Of the following, the best reason for this rule is that
  (A) it saves expense to have someone take care of simple injuries not requiring a doctor
  (B) it can keep the fire department out of a lawsuit if a simple injury is taken care of promptly
  (C) it is of great advantage that a firefighter be prepared to act quickly in an emergency
  (D) in doing first aid, it is sometimes better to do nothing than to do the wrong thing

93. The method that is *not* a recommended procedure for treating shock is
  (A) administering oxygen
  (B) placing the victim in a horizontal position
  (C) trying to prevent the loss of body heat by covering the victim with a blanket
  (D) having the victim drink warm water

94. To control severe bleeding, a tourniquet should be applied only as a last resort. Once applied, it is left in place. Of the following, the statement that is *not* correct is:
  (A) The tourniquet is placed two inches from the injury between the heart and the location of the injury.
  (B) Tighten the tourniquet only enough to stop the bleeding.
  (C) Mark the person with the letters T–K and the time of day the tourniquet was applied.
  (D) Loosen and retighten the tourniquet from time to time.

95. An open fracture generally means a fracture in which
  (A) the broken bone has penetrated the skin
  (B) the bone is broken at more than one point
  (C) the bone not only has been broken but has been dislocated as well
  (D) more than one bone has been broken

96. A newly appointed firefighter should realize that the principal function of an immediate superior officer is to
  (A) inflict discipline for rule infractions
  (B) direct the activities of subordinates so they function in a proper and acceptable manner
  (C) make recommendations to promote the most able subordinates under his or her command
  (D) keep records that properly reflect the activities of subordinates

DIRECTIONS: *Choose the correct spelling of the word to complete the sentences in questions 97 and 98.*

97. A fire's need for oxygen to continue the burning process often results in an oxygen-deficient _____ in a structural fire.
    - (A) atmosfere
    - (B) atmosphere
    - (C) atmosfear
    - (D) atmosphear
    - (E) atmospheir

98. Class A and Class B fires can be extinguished with an _____ film forming foam.
    - (A) aquius
    - (B) aqwious
    - (C) aqweus
    - (D) aqueous
    - (E) aquious

DIRECTIONS: *Choose the appropriate word or phrase to complete the sentences in questions 99 and 100.*

99. _____ they had left the fire scene, some smoldering ashes caused the fire to start up again.
    - (A) After
    - (B) While
    - (C) Until
    - (D) Whether
    - (E) Where

100. The fire was so big and lasted for such a long time that the _____ supplies were running low.
    - (A) firefighteres's
    - (B) firefighters's
    - (C) firefighters'
    - (D) firefighterz
    - (E) firefighteres

Answer Key for Practice Examination 2

1. C	26. B	51. B	76. B
2. D	27. A	52. C	77. D
3. D	28. D	53. B	78. C
4. A	29. C	54. C	79. D
5. D	30. B	55. D	80. B
6. D	31. D	56. B	81. D
7. B	32. B	57. C	82. D
8. B	33. C	58. A	83. A
9. C	34. C	59. E	84. B
10. B	35. B	60. B	85. B
11. A	36. B	61. B	86. B
12. A	37. C	62. A	87. B
13. A	38. C	63. C	88. B
14. D	39. D	64. C	89. C
15. D	40. C	65. D	90. B
16. C	41. D	66. B	91. B
17. B	42. B	67. D	92. C
18. A	43. A	68. B	93. D
19. E	44. B	69. D	94. D
20. D	45. B	70. C	95. A
21. D	46. B	71. B	96. B
22. B	47. B	72. C	97. B
23. A	48. C	73. C	98. D
24. A	49. C	74. C	99. A
25. D	50. A	75. C	100. C

Explanatory Answers for Practice Examination 2

1. **(C)** 200 Pine Street is the tallest building on the block.

2. **(D)** The Dexter Hotel at 216 Pine Street has occupants that are likely to be asleep in the early morning hours. The area is commercial/industrial, and most businesses are not open at 3 AM.

3. **(D)** This is the building farthest from the hydrant.

4. **(A)** Note the skylights on the diagram.

5. **(D)** Refer to the legend and the diagram.

6. **(D)** Refer to the legend and the diagram.

7. **(B)** 218 and 220 Pine Street are both one-story buildings.

8. **(B)** The siamese connection on the front wall of the building indicates that the building has an automatic wet sprinkler system.

9. **(C)** The direction of wind is indicated by the arrow and compass points at the top of the diagram.

10. **(B)** The building that houses Brown Lumber would require the deepest penetration of a hose stream because of the type of material stored there.

11. **(A)** Refer to the diagram and the definition of a trussed roof in the first sentence of the question.

12. **(A)** Baled material, fibers, and other materials are used in the manufacture of dresses and other clothing.

13. **(A)** The Dexter Hotel is immediately adjacent to the building afire. Fire burning through the roof of Brown Lumber would expose the hotel to flames, especially with the wind blowing from the east.

14. **(D)** Generally speaking, guests staying at a hotel are so preoccupied with their activities that they rarely take the time to familiarize themselves with the location of the fire exits and fire stairs.

15. **(D)** The firefighter knows the dangerous situation presented by the live wire and the tendency for it to recoil (wire has memory) without notice. The firefighter should protect the public until the fire department or utility company arrives.

16. **(C)** Entry into a private home or apartment is illegal unless permission to do so is given by the occupant. In the event of a fire, however, permission is not required.

17. **(B)** Bystanders in the immediate vicinity of the fire hinder effective firefighter operations. They impede movement of fire apparatus, stretching of hoselines, and rescue efforts. There is also the danger of bystanders being injured by falling debris.

18. **(A)** To square a number, just multiply the number by itself. In this case, $12 \times 12 = 144$.

19. **(E)** The amount of water available to the firefighters is the sum of all of the water in the three tankers. Thus, $1,500 + 2,500 + 3,200 = 7,200$ gallons.

20. **(D)** To find the average of a list of numbers, first add them up ($100 + 100 + 150 + 150 + 250 + 300 = 1,050$) and then divide the sum of the numbers by the number of lengths listed ($1,050/6 = 175$).

21. **(D)** To find the area of a flat surface, multiply the length times the width. $5 \times 2 = 10$ square feet

22. **(B)** The range is equal to the largest number minus the smallest number. Thus, the largest range is $135 - 50 = 85$. This is the range of the aerial ladder.

23. **(A)** The range is equal to $102 - 55 = 47$. Because $75 - 55 = 20$, $20/47 = 42.5\%$.

24. **(A)** Oxygen is a noncombustible gas; however, it will support combustion and will accelerate the fire in whatever is burning.

25. **(D)** Fires are always unpredictable. In hot, smoky conditions, the buddy system makes for greater firefighter safety.

26. **(B)** Unless the ventilation of a fire building is synchronized with advancing hoselines, the in-rush of fresh air can cause a fire to accelerate.

27. **(A)** Patrons in a theater generally are surprised when the lights are turned on suddenly. The manager calmed the patrons at once and informed them of how to make a safe and speedy exit. This was very well done.

28. **(D)** A sprinkler head has a fusible metal disc, which is used to hold back the water. The heat from the fire causes the disc to melt at a predetermined temperature and, in doing so, releases the water. It cannot be used again; therefore, it must be replaced.

29. **(C)** The firefighter's job is to extinguish fires, to save lives, and at every opportunity, to make an effort to hold property damage from fire or water to a minimum. With the fire out and no danger to life, the firefighter correctly saved property.

30. **(B)** The firefighter is on duty to respond to fires and emergencies. If he "kept an eye" on the child, he would not be available to respond with his unit.

31. **(D)** The baking soda would smother the fire. The other extinguishing agents would cause the grease to splatter and would spread the fire from the pan to surrounding areas.

32. **(B)** Before obtaining a certificate or permit from the fire department, people applying must show fire department examiners that they are trained and qualified to engage in the particular hazardous work.

33. **(C)** By reporting the complaint, the officer will then cause an official inspection to be made. If the doorway is blocked, a violation order will be issued to the person in charge.

34. **(C)** The bankrupt owner of the store deliberately set the fire in the store. The other three fires were started accidentally.

35. **(B)** The two-way radio enables the fire company to be away from the fire station and still be in radio contact with the fire alarm dispatcher and available to respond to alarms.

36. **(B)** Oxyacetylene torches generate a high intensity flame that is used to cut metal. As the metal is cut, the hot pieces and sparks bounce from the metal. Numerous fires are started by sparks and hot metal falling on combustible material.

37. **(C)** The firefighter in uniform serves as a beacon to a person seeking aid. The uniform also identifies the firefighter as a person with authority. This may be a deterrent to anyone intent on committing a criminal act.

38. **(C)** The majority of wood and wood products at an outdoor lumberyard are stored outdoors and are exposed to all weather conditions. Wetting by hose streams should have little adverse effect on the lumber.

39. **(D)** The arsonist generally will open windows to allow air in to accelerate the fire. An open or unlocked door usually indicates that someone was in the building prior to the arrival of the firefighters.

40. **(C)** Fire prevention is a very important aspect of the firefighter's role. Study of the causes of fires can lead to steps to eliminate those causes at other locations, thus preventing future fires.

41. **(D)** The term "fireproof building" means that the walls, floors, stairways, partitions, and other structural components of the building are made of noncombustible material and will resist fire. The contents and furnishings, however, are combustible and are likely to burn.

42. **(B)** The fire department would investigate a fire for the reasons stated in (A), (C), and (D). The amount of loss for insurance purposes would be investigated by an adjuster hired by the insurance company or the victim of the fire.

43. **(A)** The lieutenant might have been aware of other circumstances or conditions at the fire that the probationary firefighter, because of inexperience, was not aware of. The officer who gave the order is in the best position to explain it.

44. **(B)** Pier fires usually are fires of long duration that require enormous quantities of water to extinguish. By using the salt water, we have an unlimited supply and we can conserve our fresh-water resources.

45. **(B)** Carbon monoxide is a toxic gas. The atmospheric air we breathe is composed of 84 percent nitrogen and 16 percent oxygen; obviously, nitrogen is not a toxic gas.

46. **(B)** The flame indicates that there is sufficient oxygen. Gas concentration is low enough to be handled by the mask.

47. **(B)** This is stated in the last sentence of the paragraph.

48. **(C)** The canister-type gas mask contains chemicals that filter toxic gases and smoke, resulting in breathable air.

49. **(C)** This is stated in the first sentence.

50. **(A)** This is stated in the second sentence. Doorway openings permitted in the wall are a maximum of 25 percent of 22 feet, or 5.5 feet. One opening would be 3 feet, and two openings would be a total of 6 feet. Therefore, (A) must be the answer.

51. **(B)** This is stated in the last sentence. The maximum opening for a motor vehicle is 140 square feet. The opening in the diagram is 13 feet × 11 feet, or 143 square feet. Therefore, it is too large.

52. **(C)** The purpose of fire department inspections of any and all property is to eliminate conditions creating undue fire hazards.

53. **(B)** The average household toaster uses 1400 watts of energy, much more than the other items mentioned in (A), (C), and (D).

54. **(C)** Choices (A), (B), and (D) are programs that investigate the causes of water usage, as stated in the first sentence.

55. **(D)** This is stated in the last sentence.

56. **(B)** To obtain the amount used by each person, take the total number of gallons consumed and divide this number by the total number of people (population).

57. **(C)** This entire passage discusses the purpose and functions of the Incident Command System (ICS).

58. **(A)** Under the heading of command, the passage states that, for a large-scale operation to run safely and effectively, it is necessary to have people in charge.

59. **(E)** Under the heading of finance, the passage states that finance is usually only a concern during large-scale incidents that last for a particularly long period of time.

60. **(B)** The only thing that is not listed as a responsibility of logistics is rescue.

61. **(B)** The first paragraph states that, if an investigator is not available immediately after a fire, it is important for the firefighters on the scene to ensure proper collection of the evidence. The second paragraph then states that the firefighters should tag, mark, and photograph the evidence as soon as possible.

62. **(A)** The third paragraph indicates that only the person's name, time of entry, time of departure, and a description of the items taken should be logged.

63. **(C)** The first sentence of the second paragraph states that the fire department has the authority to control a fire scene for as long after a fire as necessary.

64. **(C)** The last sentence of the second paragraph states that it is important for firefighters to not damage any evidence.

65. **(D)** Friction opposes motion. By placing a coarse substance between two smooth surfaces, friction is generated and skidding is reduced.

66. **(B)** Heated air expands; it becomes lighter and rises. This is called convection.

67. **(D)** The ladder with the gentler grade is more stable and is less likely to tip over. The horizontal distance between the wall and the base of the ladder is determined by dividing the length of the ladder by 5 and adding 2. Example: For a 20-foot ladder, dividing by 5 equals 4. 4 plus 2 equals 6. 6 feet is the proper distance from the wall for the 20-foot ladder.

68. **(B)** Spontaneous heating is the increase in temperature of a material or substance without drawing heat from its surroundings.

69. **(D)** Food that has been subjected to unusual heat and perhaps also to water might well have undergone changes that render it unfit for human consumption. For the protection of the public, it is important for this possibility to be recognized and for the food to be inspected.

70. **(C)** The most common method of extinguishing a fire is to pour water on it. Water is a very effective cooling agent. The action of water is to reduce the temperature of the burning material.

71. **(B)** The principle that applies to this problem is as follows: Pressure in a liquid is equal to the height multiplied by the density, or P = H × D. Because siphon 2 has a greater depth, the liquid at the bottom of the siphon 2 tube has a greater pressure than that of siphon 1. The greater pressure also produces greater velocity of the moving liquid. Therefore, the flow in siphon 2 would be faster, making it more efficient.

72. **(C)** Static electricity is the electrification of material through physical contact and separation. Sparks might result, constituting a fire or explosion hazard.

73. **(C)** Salt water has greater density than fresh water. For this reason, it is easier to float in salt water than in fresh water.

74. **(C)** The inclined ramp (or plane) is simply a sloping platform that enables a person to raise an object without having to lift it vertically. It is a simple machine that helps you perform work.

75. **(C)** A distinct mechanical advantage is gained by using a large lever (the crank) to move a small object (the drum).

76. **(B)** Both balls will reach the ground at the same time because the gravitational force on objects with the same surface size varies with the distance of the objects from the center of the earth. The weight of the objects does not matter.

77. **(D)** If you do not know the answer, you can use reasoning to figure out this question. Choices (B) and (C) are irrelevant. Choice (A) makes no sense. It is reasonable to assume that the space in a gasoline drum is filled with gasoline vapors.

78. **(C)** The explanation is based on a well-known principle of centrifugal force.

79. **(D)** When water freezes, it increases its volume by one-eleventh. When 11 cubic inches of water freeze, for example, 12 cubic inches of ice form. This is why water pipes burst and auto radiators without antifreeze are severely damaged.

80. **(B)** The chart indicates that Class C fires involve energized electrical equipment. Because water is a good conductor of electricity, there is a serious risk of electrical shock if water is used on a Class C fire.

81. **(D)** The table indicates that, when paints are the combustible material, oxygen exclusion (the blanketing effect) is the best way to extinguish the fire.

82. **(D)** When resistance is placed against the drill bit, the tool will function satisfactorily and the hole will be made. As the bit approaches the underside of the wood, the resistance is lessened and the bit is forced through the wood, splintering the underside.

83. **(A)** Jumping into a net is not an activity with which many people have had experience. The person jumping might be unable to gauge distance accurately. Firefighters must be alert to position the net under the person.

84. **(B)** The metal spoon is a good conductor and will rapidly conduct heat away from the water. This reduces the amount of heat absorbed by the glass and prevents it from breaking.

85. **(B)** Friction in a pipe or a hose varies directly with the length. The longer the pipe or hose, the greater the friction. The immediate result of friction in a hose is to cut down the available pressure at the nozzle.

86. **(B)** This is a principle of pressure in fluids. Pressure applied to a confined fluid from within is transmitted in all directions without differentiation. Therefore, all points in the 100-foot pipe would have the same pressure.

87. **(B)** This is an example of a second-class lever. (The weight is between the fulcrum and the lift.) By moving the weight closer to the fulcrum, an increase in distance between the load and the lift point is achieved. This increases the mechanical advantage of the lever, and less effort is necessary to lift the load.

88. **(B)** This is stated in the third sentence. Alterations in old buildings frequently create voids or channels that allow fire to extend throughout a building. Removal of partitions and/or structural members have resulted in the collapse of buildings.

89. **(C)** This is the meaning of the fourth sentence.

90. **(B)** This is stated in the last sentence.

91. **(B)** See the third sentence.

92. **(C)** Many times, the firefighter is the first person at the scene of an accident. By being knowledgeable in first aid, he or she can give prompt and efficient care to an injured person. This is essential when attending to life-threatening emergencies.

93. **(D)** Proper treatment of shock is to avoid rough handling, control bleeding, assure breathing, and give nothing by mouth.

94. **(D)** Loosening the tourniquet can result in death by causing additional bleeding, by dislodging clots, and by causing tourniquet shock.

95. **(A)** An open fracture, as the name implies, is a fracture in which there is an open wound. Prompt and efficient care should be given to a closed fracture to prevent it from becoming an open or compound fracture.

96. **(B)** The primary objective of the supervisor is to get the job accomplished. Directing is instructing, informing, and ordering subordinates about what to do and sometimes how to do it to achieve the objectives of the department.

97. **(B)** The correct spelling is "atmosphere."

98. **(D)** The correct spelling is "aqueous."

99. **(A)** The word "after" is the best word to complete the sentence.

100. **(C)** The word "firefighters'" is the best word to complete the sentence. The apostrophe after the "s" in "firefighters'" indicates that the supplies belong to multiple firefighters.

Answer Sheet for Practice Examination 3

1. Ⓐ Ⓑ Ⓒ Ⓓ Ⓔ
2. Ⓐ Ⓑ Ⓒ Ⓓ Ⓔ
3. Ⓐ Ⓑ Ⓒ Ⓓ Ⓔ
4. Ⓐ Ⓑ Ⓒ Ⓓ Ⓔ
5. Ⓐ Ⓑ Ⓒ Ⓓ Ⓔ
6. Ⓐ Ⓑ Ⓒ Ⓓ Ⓔ
7. Ⓐ Ⓑ Ⓒ Ⓓ Ⓔ
8. Ⓐ Ⓑ Ⓒ Ⓓ Ⓔ
9. Ⓐ Ⓑ Ⓒ Ⓓ Ⓔ
10. Ⓐ Ⓑ Ⓒ Ⓓ Ⓔ
11. Ⓐ Ⓑ Ⓒ Ⓓ Ⓔ
12. Ⓐ Ⓑ Ⓒ Ⓓ Ⓔ
13. Ⓐ Ⓑ Ⓒ Ⓓ Ⓔ
14. Ⓐ Ⓑ Ⓒ Ⓓ Ⓔ
15. Ⓐ Ⓑ Ⓒ Ⓓ Ⓔ
16. Ⓐ Ⓑ Ⓒ Ⓓ Ⓔ
17. Ⓐ Ⓑ Ⓒ Ⓓ Ⓔ
18. Ⓐ Ⓑ Ⓒ Ⓓ Ⓔ
19. Ⓐ Ⓑ Ⓒ Ⓓ Ⓔ
20. Ⓐ Ⓑ Ⓒ Ⓓ Ⓔ
21. Ⓐ Ⓑ Ⓒ Ⓓ Ⓔ
22. Ⓐ Ⓑ Ⓒ Ⓓ Ⓔ
23. Ⓐ Ⓑ Ⓒ Ⓓ Ⓔ
24. Ⓐ Ⓑ Ⓒ Ⓓ Ⓔ
25. Ⓐ Ⓑ Ⓒ Ⓓ Ⓔ
26. Ⓐ Ⓑ Ⓒ Ⓓ Ⓔ
27. Ⓐ Ⓑ Ⓒ Ⓓ Ⓔ
28. Ⓐ Ⓑ Ⓒ Ⓓ Ⓔ
29. Ⓐ Ⓑ Ⓒ Ⓓ Ⓔ
30. Ⓐ Ⓑ Ⓒ Ⓓ Ⓔ
31. Ⓐ Ⓑ Ⓒ Ⓓ Ⓔ
32. Ⓐ Ⓑ Ⓒ Ⓓ Ⓔ
33. Ⓐ Ⓑ Ⓒ Ⓓ Ⓔ
34. Ⓐ Ⓑ Ⓒ Ⓓ Ⓔ
35. Ⓐ Ⓑ Ⓒ Ⓓ Ⓔ
36. Ⓐ Ⓑ Ⓒ Ⓓ Ⓔ
37. Ⓐ Ⓑ Ⓒ Ⓓ Ⓔ
38. Ⓐ Ⓑ Ⓒ Ⓓ Ⓔ
39. Ⓐ Ⓑ Ⓒ Ⓓ Ⓔ
40. Ⓐ Ⓑ Ⓒ Ⓓ Ⓔ
41. Ⓐ Ⓑ Ⓒ Ⓓ Ⓔ
42. Ⓐ Ⓑ Ⓒ Ⓓ Ⓔ
43. Ⓐ Ⓑ Ⓒ Ⓓ Ⓔ
44. Ⓐ Ⓑ Ⓒ Ⓓ Ⓔ
45. Ⓐ Ⓑ Ⓒ Ⓓ Ⓔ
46. Ⓐ Ⓑ Ⓒ Ⓓ Ⓔ
47. Ⓐ Ⓑ Ⓒ Ⓓ Ⓔ
48. Ⓐ Ⓑ Ⓒ Ⓓ Ⓔ
49. Ⓐ Ⓑ Ⓒ Ⓓ Ⓔ
50. Ⓐ Ⓑ Ⓒ Ⓓ Ⓔ
51. Ⓐ Ⓑ Ⓒ Ⓓ Ⓔ
52. Ⓐ Ⓑ Ⓒ Ⓓ Ⓔ
53. Ⓐ Ⓑ Ⓒ Ⓓ Ⓔ
54. Ⓐ Ⓑ Ⓒ Ⓓ Ⓔ
55. Ⓐ Ⓑ Ⓒ Ⓓ Ⓔ
56. Ⓐ Ⓑ Ⓒ Ⓓ Ⓔ
57. Ⓐ Ⓑ Ⓒ Ⓓ Ⓔ
58. Ⓐ Ⓑ Ⓒ Ⓓ Ⓔ
59. Ⓐ Ⓑ Ⓒ Ⓓ Ⓔ
60. Ⓐ Ⓑ Ⓒ Ⓓ Ⓔ
61. Ⓐ Ⓑ Ⓒ Ⓓ Ⓔ
62. Ⓐ Ⓑ Ⓒ Ⓓ Ⓔ
63. Ⓐ Ⓑ Ⓒ Ⓓ Ⓔ
64. Ⓐ Ⓑ Ⓒ Ⓓ Ⓔ
65. Ⓐ Ⓑ Ⓒ Ⓓ Ⓔ
66. Ⓐ Ⓑ Ⓒ Ⓓ Ⓔ
67. Ⓐ Ⓑ Ⓒ Ⓓ Ⓔ
68. Ⓐ Ⓑ Ⓒ Ⓓ Ⓔ
69. Ⓐ Ⓑ Ⓒ Ⓓ Ⓔ
70. Ⓐ Ⓑ Ⓒ Ⓓ Ⓔ
71. Ⓐ Ⓑ Ⓒ Ⓓ Ⓔ
72. Ⓐ Ⓑ Ⓒ Ⓓ Ⓔ
73. Ⓐ Ⓑ Ⓒ Ⓓ Ⓔ
74. Ⓐ Ⓑ Ⓒ Ⓓ Ⓔ
75. Ⓐ Ⓑ Ⓒ Ⓓ Ⓔ
76. Ⓐ Ⓑ Ⓒ Ⓓ Ⓔ
77. Ⓐ Ⓑ Ⓒ Ⓓ Ⓔ
78. Ⓐ Ⓑ Ⓒ Ⓓ Ⓔ
79. Ⓐ Ⓑ Ⓒ Ⓓ Ⓔ
80. Ⓐ Ⓑ Ⓒ Ⓓ Ⓔ
81. Ⓐ Ⓑ Ⓒ Ⓓ Ⓔ
82. Ⓐ Ⓑ Ⓒ Ⓓ Ⓔ
83. Ⓐ Ⓑ Ⓒ Ⓓ Ⓔ
84. Ⓐ Ⓑ Ⓒ Ⓓ Ⓔ
85. Ⓐ Ⓑ Ⓒ Ⓓ Ⓔ
86. Ⓐ Ⓑ Ⓒ Ⓓ Ⓔ
87. Ⓐ Ⓑ Ⓒ Ⓓ Ⓔ
88. Ⓐ Ⓑ Ⓒ Ⓓ Ⓔ
89. Ⓐ Ⓑ Ⓒ Ⓓ Ⓔ
90. Ⓐ Ⓑ Ⓒ Ⓓ Ⓔ
91. Ⓐ Ⓑ Ⓒ Ⓓ Ⓔ
92. Ⓐ Ⓑ Ⓒ Ⓓ Ⓔ
93. Ⓐ Ⓑ Ⓒ Ⓓ Ⓔ
94. Ⓐ Ⓑ Ⓒ Ⓓ Ⓔ
95. Ⓐ Ⓑ Ⓒ Ⓓ Ⓔ
96. Ⓐ Ⓑ Ⓒ Ⓓ Ⓔ
97. Ⓐ Ⓑ Ⓒ Ⓓ Ⓔ
98. Ⓐ Ⓑ Ⓒ Ⓓ Ⓔ
99. Ⓐ Ⓑ Ⓒ Ⓓ Ⓔ
100. Ⓐ Ⓑ Ⓒ Ⓓ Ⓔ

TEAR HERE

PRACTICE EXAMINATION 3

The time allowed for the entire examination is 3½ hours.

DIRECTIONS: *Each question has four suggested answers lettered A, B, C, and D. Decide which one is the best answer. On the sample answer sheet, find the question number that corresponds to the answer you have selected and darken the area with a soft pencil.*

Memory Question

The following passage describes an incident encountered by firefighters while engaging in their day-to-day activities. You will have five minutes to study this passage. Questions 1 through 15 are based on the details described in these paragraphs. Therefore, you should make every effort to memorize the highlights as they are described during the time allowed for memorization.

During the summer of 1981, a severe water shortage existed that was the result of an extended heat wave and drought. Hydrants opened by citizens in an effort to gain relief from the heat depleted water resources rapidly, and emergency measures were undertaken by the city to conserve water.

Fire department personnel conducted a survey that revealed that 600 hydrants were opened daily by citizens and an average of 1,100 gallons of water per minute were wasted at each open hydrant. Hydrant patrols were implemented using 75 engine companies on two-hour shifts to close down open hydrants.

One day, while on hydrant patrol, Engine 299 stopped to shut down an open hydrant that was being used by area residents to gain relief from the summer heat. While the firefighters were in the act of closing the hydrant, they were subjected to severe verbal abuse by the crowd that quickly formed around them. A potentially explosive situation developed rapidly. The firefighters closed the hydrant and quickly left the scene to return to the firehouse.

They were in the firehouse for a very short time when an alarm was received to return to the same street for a reported fire in an occupied multiple-dwelling building. Engine 299 responded to the alarm. As they turned into the street, they were greeted with a barrage of rocks and bottles thrown from doorways and rooftops. The officer in command ordered the firefighters to take cover immediately and transmitted a radio message requesting police assistance. Police officers arrived and dispersed the crowd. The firefighters then checked the location of the reported fire and learned that there was no actual fire; it was a false alarm. They returned to the firehouse, and a report of the incident was forwarded to the fire commissioner.

The following day, the deputy chief requested that a meeting be arranged between fire department officials and community leaders in an effort to discuss and resolve the problem. At the meeting, fire department representatives informed community leaders of the importance of having adequate water pressure available to extinguish fires and attempted to convince residents that firefighters were there to help them. Residents also were informed that sprinkler caps for fire hydrants were available at the firehouse. The use of these caps would result in the saving of much water. It then would be possible for the residents to use the hydrants to cool themselves, and sufficient water pressure would be maintained to fight fires.

Community leaders must be convinced that responding to false alarms can result in the destruction of property and the deaths of citizens because firefighters are not able to respond to an actual fire if they are responding to a false alarm.

DIRECTIONS: *Questions 1 through 15 relate to the information in the preceding passage and should be answered in accordance with this information.*

1. According to the passage, a severe water shortage existed in the city during the summer of 1981. This shortage was due to
    (A) broken water mains
    (B) open hydrants
    (C) an extended heat wave and drought
    (D) misuse of water by industry

2. The firefighters of Engine 299 stopped to shut down the open hydrant while
    (A) returning to quarters
    (B) responding to an alarm
    (C) on alarm box inspection duty
    (D) on hydrant patrol duty

3. From the information given in the passage, we can conclude that the main concern of the fire department was
    (A) to provide the area residents with sprinkler caps
    (B) to have water available for firefighting
    (C) education of the residents in the area
    (D) to show the cooperation between the police and fire departments

4. If a lieutenant and five firefighters are on duty in each engine company, the total personnel hours spent by the fire department on hydrant patrol for one day was
    (A) 150 hours
    (B) 350 hours
    (C) 600 hours
    (D) 900 hours

5. The *most important* factor the fire officials wanted to convey to the community residents at the meeting was
    (A) to conserve water
    (B) to use sprinkler caps whenever possible
    (C) that false alarms can result in death
    (D) not to abuse firefighters while they are on duty

6. During extended periods of hot weather, hydrant patrol duty usually becomes a daily routine. Of the following, the action that should *not* be taken during these patrols is to
    (A) learn the location and the condition of the hydrants in the district
    (B) use force on area residents when encountering opposition to the firefighters
    (C) train chauffeurs to operate the fire apparatus
    (D) shut down open hydrants to ensure adequate water pressure

7. The officer in command ordered the firefighters to take shelter from the barrage of rocks and bottles being thrown. The lieutenant's actions were
   (A) improper; the lieutenant should have first ordered a search for the fire
   (B) proper; however, the lieutenant should have tried to calm the crowd first
   (C) improper; the lieutenant should have requested more fire department units to respond to the scene
   (D) proper; the lieutenant's immediate responsibility was the safety and protection of the firefighters under attack

8. The firefighters shut down the hydrant and hastily left the scene in order to
   (A) respond to an alarm for a fire in a multiple dwelling
   (B) respond to a false alarm
   (C) obtain police assistance
   (D) avoid a physical confrontation with the crowd

9. Fire department engine companies were deployed to shut down open hydrants. The total number of hours on patrol duty by fire department companies was
   (A) 150 hours
   (B) 200 hours
   (C) 250 hours
   (D) 300 hours

10. It would be incorrect to state that the police were called to the scene
    (A) to shut down the open hydrant
    (B) to control the crowd
    (C) for protection of the firefighters
    (D) by the lieutenant in command of Engine 299

11. Open hydrants were depleting the water supply rapidly. In a two-hour period, an open hydrant was discharging approximately
    (A) 1,100 gallons of water
    (B) 2,200 gallons of water
    (C) 66,000 gallons of water
    (D) 132,000 gallons of water

12. According to the information contained in the passage, Engine 299 responded from headquarters to
    (A) investigate a complaint of rubbish on a roof
    (B) a reported fire in an occupied private dwelling
    (C) a reported fire in a multiple dwelling
    (D) assist police at a false alarm

13. The average number of hydrants shut down each day by each engine company was
    (A) 4
    (B) 6
    (C) 8
    (D) 10

14. The officer in command of Engine 299 forwarded a report of the rock- and bottle-throwing incident. This report was forwarded to the
    (A) fire commissioner
    (B) deputy chief
    (C) community leaders
    (D) police department

15. It would be reasonable to believe that the main purpose of the meeting between fire officials and the community leaders was to
    (A) inform the residents of the importance of the firefighters
    (B) stress the importance of having adequate water pressure available for firefighting
    (C) remind the community that the firefighter is their friend
    (D) make the community aware of the sprinkler caps

16. The expansion ratio for medium-expansion foams are from 50:1 to 300:1. If three gallons of expansion concentrate are used, how many gallons of foam will be produced?
    (A) from 150 to 500 gallons
    (B) from 150 to 900 gallons
    (C) from 100 to 600 gallons
    (D) from 300 to 900 gallons
    (E) from 200 to 950 gallons

17. A fire consumed portions of three floors of a building. On the first floor, 1,300 square feet were damaged; on the second floor, 12,000 square feet were damaged; on the third floor, 7,500 square feet were damaged. What is the total damaged area of the building?
    (A) 13,300 square feet
    (B) 19,500 square feet
    (C) 20,800 square feet
    (D) 23,500 square feet
    (E) 25,300 square feet

18. If a water tanker can pump out water at 10,000 gallons per hour, how long will it take to pump out 1,000 gallons of water?
    (A) 6 minutes
    (B) 9 minutes
    (C) 10 minutes
    (D) 12 minutes
    (E) 15 minutes

19. In an enclosed room, a house fire can create temperatures of 900°F. How many times hotter is this than an average room temperature of 75°F?
    (A) 10
    (B) 12
    (C) 13
    (D) 15
    (E) 20

20. If a fully extended 24-foot ladder is needed to reach a second-story window, how long should a ladder be to reach a fifth-story window?
    (A) 35 feet
    (B) 40 feet
    (C) 50 feet
    (D) 60 feet
    (E) 65 feet

21. In any city or suburb with a population of at least 20,000 people, there should be two fire hydrants for every square city block. If there are 75 square city blocks in one mile, how many fire hydrants should there be in two square miles?
    (A) 150
    (B) 200
    (C) 225
    (D) 250
    (E) 300

DIRECTIONS: *Choose the correct spelling of the word to complete the sentences in questions 22 and 23.*

22. There are _____ advantages to keeping equipment and protective clothing clean and free from defects.
    (A) numerous
    (B) numerus
    (C) numirous
    (D) numourus
    (E) numirus

23. _____ rope often is used instead of natural rope in rescue operations.
    (A) Sinthetic
    (B) Synthettic
    (C) Synthetick
    (D) Sinthetik
    (E) Synthetic

DIRECTIONS: *Choose the appropriate word or phrase to complete the sentences in questions 24 and 25.*

24. Firefighter Campbell _____ a captain at station 32.
    (A) am
    (B) is
    (C) are
    (D) were
    (E) being

25. Since the fire was over, Mr. Cook had _____ to collect what was left of his belongings.
    (A) began
    (B) begin
    (C) beginning
    (D) begins
    (E) begun

26. In New York, schools are required to hold at least 12 fire drills each school year, 8 of which must be conducted between September and December. Of the following, the best justification for holding most fire drills during the September 1 to December 1 period is that
    (A) most fires occur during that period
    (B) pupils are trained in evacuation of their schools at the beginning of the school year
    (C) the weather is milder during that period
    (D) school attendance is higher at the beginning of the school year

27. Following a severe smoky fire in the lobby of a hotel, a dead body was found in a room on the eighth floor. The coroner who examined the body found that the victim died from the inhalation of toxic gases and smoke. The most probable explanation of the occurrence is that
    (A) the coroner made an error in the findings
    (B) the victim inhaled the gases and smoke in the lobby, fled to the eighth floor to escape the fire, and collapsed there
    (C) the gases and smoke traveled through vertical openings to the upper stories where they were inhaled by the victim.
    (D) a second fire occurred simultaneously in the upper stories of the hotel

28. The basic assumption of fire prevention educational programs is that people frequently
    (A) must be forced into obeying fire laws
    (B) are unaware of the dangers involved in some of their actions
    (C) do not care whether their actions are dangerous
    (D) assume that fire insurance protects them against all fire loss

29. The statement that is most accurate is that visibility in buildings on fire
    (A) is not a serious problem mainly because fires give off sufficient light to enable firefighters to see clearly
    (B) is a serious problem mainly because fires often knock out the electrical system of the building
    (C) is not a serious problem mainly because most fires occur during daylight hours
    (D) is a serious problem mainly because smoky conditions are often encountered

30. While operating at a five-alarm fire, a firefighter is approached by an obviously intoxicated man who claims to be a former firefighter and who offers to help put out the fire. In this situation, the best of the following courses of action for the firefighter is to
    (A) give the man some easy-to-perform task
    (B) refer the man to the officer in command of the fire
    (C) decline the offer of help and ask the man to remain outside the fire lines
    (D) ask to see the man's credentials and, if he is a former firefighter as he claims, put him to work stretching hoselines

31. A civilian came into a firehouse to report that some adolescents had turned on a hydrant in the neighborhood. After getting the location of the hydrant, the firefighter on housewatch asked the civilian for his name and was told "Smith." The firefighter then thanked the civilian and proceeded to process the report. The firefighter's actions in this situation were
    (A) proper
    (B) proper, except that the firefighter should not have wasted time asking for the civilian's name
    (C) proper, except that he should have found out whether the civilian's name really was Smith
    (D) proper, except that the firefighter should have obtained the civilian's first name and address as well as his family name

32. Fire apparatus are not permitted to use their sirens when returning to headquarters. Of the following, the main justification for this restriction is that
    (A) the chances of being involved in traffic accidents are reduced
    (B) there is no need for prompt return to quarters after a fire has been extinguished
    (C) apparatus return to quarters using less crowded streets than when responding to alarms
    (D) the officer in command is better able to exchange radio messages with the fire alarm dispatcher

33. A fire escape platform is attached to a building flush with the window sill leading onto it. The most important reason for this arrangement is that it reduces the chance that
    (A) intruders will utilize the fire escapes for illegal purposes because they will be visible from inside the apartment
    (B) tenants will place garbage or other obstructions on the platform because these objects will be visible from inside the apartment
    (C) persons using the fire escape in an emergency will trip as they leave the apartment
    (D) sagging or loosening of the supports will occur without coming to the attention of persons occupying the apartment

34. A civilian, on his way home from work one evening, hears an alarm ringing and sees water running out of a sprinkler discharge pipe on the side of a building. No smoke or other indications of fire are seen. There is a padlock on the entrance to the building, no fire alarm box is in sight, and a firehouse is located $1\frac{1}{2}$ blocks away. In this situation, it would be most proper for the civilian to
    (A) take no action because there is no evidence of a fire or other emergency
    (B) attempt to ascertain the name of the building's owner and notify him of the situation
    (C) send an alarm from the closest fire alarm box
    (D) go to the firehouse and inform the firefighter at the desk of the situation

35. Running a hoseline between two rungs of a ladder into a building on fire generally is considered to be a
    (A) good practice, mainly because resting the hose on the rungs of the ladder reduces the load that must be carried by the firefighters
    (B) poor practice, mainly because the maneuverability of the ladder is reduced
    (C) good practice, mainly because the weight of the hoseline holds the ladder more securely against the side of the building
    (D) poor practice, mainly because the ladder can be damaged by the additional weight of the hoseline filled with water

36. Suppose you are the firefighter on housewatch duty when a woman rushes into the firehouse and reports that a teenage gang is assaulting a man on the street several blocks away. In this situation, the best of the following courses of action for you to take first is to
    (A) notify the police department of the reported incident
    (B) go to the scene of the disturbance and verify the woman's report
    (C) suggest to the woman that she report the matter to the police department
    (D) notify your fellow firefighters of the incident and all go to the aid of the man being assaulted

37. A firefighter on the way to work is standing on a subway platform waiting for a train. Suddenly, a man standing nearby collapses to the floor, showing no sign of breathing. In this situation, the firefighter should
    (A) run to the nearest telephone and summon an ambulance
    (B) designate a responsible person to look after the victim and then continue on to the firehouse
    (C) administer first-aid measures to restore breathing
    (D) take no action because the victim is obviously dead

38. Fire companies often practice methods of firefighting and lifesaving in public places. The best justification for this practice is that it
    (A) provides facilities for practicing essential operations that are not readily available in fire department installations
    (B) shows the taxpaying public that they are getting their money's worth
    (C) attracts youngsters and interests them in firefighting as a career
    (D) makes certain that all company equipment is in the best possible working condition

39. The fire that generally would present the greatest danger to surrounding buildings from flying sparks or brands is a fire in a(an)
    (A) warehouse storing paper products
    (B) factory manufacturing plastics
    (C) outdoor lumberyard
    (D) open-type parking garage

40. A fire marshal, questioning firefighters at the scene of a suspicious fire, obtains some conflicting statements about details of the fire situation present upon their arrival. The most likely explanation of this conflict is that
    (A) the firefighters have not been properly trained to carry out their part in arson investigations
    (B) the fire marshal did not give the firefighters time to collect and organize their impressions of the fire scene
    (C) witnesses to an event see the situation from their own point of view and seldom agree on all details
    (D) details are of little importance if there is general agreement on major matters

41. A firefighter on duty at a theater who discovers standees obstructing aisles should immediately
    (A) report the situation to his or her superior officer
    (B) order the standees to move out of the aisles
    (C) ask the theater manager to correct the situation
    (D) issue a summons to the usher assigned to that area

42. At a fire on the fourth floor of an apartment house, the first engine company to arrive advanced a hoseline up the stairway to the third floor before charging the hose with water. The main reason that the firefighters delayed charging their line is that an empty line
(A) is less likely to whip about and injure firefighters
(B) is easier to carry
(C) won't leak water
(D) is less subject to damage

43. Suppose the owner of a burning tenement building complains that, although the fire is located on the first floor, firefighters are chopping holes in the roof. Of the following, the most appropriate reason you can give for their action is that the fire can be fought most effectively by permitting
(A) smoke and hot gases to escape
(B) firefighters to attack the fire from above
(C) firefighters to gain access to the building through the holes
(D) immediate inspection of the roof for extension of the fire

44. The fire department always endeavors to purchase the best apparatus and equipment and to maintain it in the best condition. The main justification for this policy is that
(A) public confidence in the department is increased
(B) failure of equipment at a fire can have serious consequences
(C) replacement of worn out parts often is difficult
(D) the dollar cost to the department is less in the long run

45. The statement about smoke that is most accurate is that smoke is
(A) irritating but not dangerous in itself
(B) irritating and dangerous only because it can reduce the oxygen content of the air breathed
(C) dangerous because it can reduce the oxygen content of the air breathed and often contains toxic gases
(D) dangerous because it supports combustion

46. Suppose you are a firefighter making a routine inspection of a rubber goods factory. During the inspection, you discover some minor violations of the Fire Prevention Code. When you call these violations to the attention of the factory owner, he becomes annoyed and tells you that he is the personal friend of high officials in the fire department and city government. Under these circumstances, the best course for you to follow is to
(A) summon a police officer to arrest the owner for attempting to intimidate a public official performing a duty
(B) make a very thorough inspection and serve summonses for every possible violation of the Fire Department Code
(C) ignore the owner's remarks and continue the inspection in your usual manner
(D) try to obtain from the owner the names and positions of his friends

47. It has been suggested that property owners should be charged a fee each time the fire department is called to extinguish a fire on their property. Of the following, the best reason for rejecting this proposal is that
(A) delays in calling the fire department might result
(B) many property owners don't occupy the property they own
(C) property owners might resent such a charge because they pay real estate taxes
(D) it might be difficult to determine on whose property a fire started

48. Standpipe systems of many bridges are the dry pipe type. A dry pipe system has no water in the pipes when not in use. When water is required, it is necessary first to pump water into the system. The main reason for using a dry standpipe system is to prevent
    (A) corrosion of the pipes
    (B) freezing of water in the pipes
    (C) waste of water through leakage
    (D) strain on the pumps

49. Assume you are a firefighter on your way home after completing your tour of duty. Just as you are about to enter the subway, a man runs up to you and reports a fire in a house located five blocks away. You recognize the man as a mildly retarded but harmless person who frequently loiters around firehouses and at fires. Of the following, the best action for you to take is to
    (A) run to the house to see if there really is a fire
    (B) call in an alarm from a nearby telephone
    (C) ignore the report because of the man's mental condition
    (D) call the police and ask that a radio patrol car investigate the report

DIRECTIONS: *Use the information in the following passage to answer question 50.*

Rescue operations often take place in confined areas that do not allow for natural or forced ventilation. Due to the dangers that these types of situations present to a firefighter, it is extremely important to use a system of communication so that the firefighter in the confined space can signal to other firefighters if there is trouble. A method of signaling called the OATH method has been developed for this type of situation. The letter O indicates that the individual is okay, A indicates that the individual wants to advance, T indicates the individual wants to be taken up, and H indicates that the individual needs help. One tug on a safety line stands for O, two tugs for A, three tugs for T, and four tugs for H.

50. If a firefighter is working in a deep, narrow well and wants to be taken back up to the surface, how many times should he tug on the lifeline?
    (A) 1
    (B) 2
    (C) 3
    (D) 4
    (E) 5

DIRECTIONS: *Answer questions 51 through 55 on the basis of the information found in the following passage.*

1. The sizes of living rooms shall meet the following requirements:
    a. In each apartment, there shall be at least one living room containing at least 120 square feet of clear floor area; every other living room, except a kitchen, shall contain at least 70 square feet of clear floor area.
    b. Every living room that contains less than 80 square feet of clear floor area or that is located in the cellar or basement shall be at least 9 feet high; every other living room shall be at least 8 feet high.

2. Apartments containing three or more rooms can have dining bays, which shall not exceed 55 square feet in floor surface area and shall not be deemed separate rooms or subject to the requirements for separate rooms. Every such dining bay shall be provided with at least one window containing an area at least one-eighth of the floor surface area of such dining bay.

51. The minimum volume of a living room, other than a kitchen, that meets the minimum requirements listed in the passage is one that measures
    (A) 70 cubic feet
    (B) 80 cubic feet
    (C) 630 cubic feet
    (D) 640 cubic feet

52. A builder proposes to construct an apartment house containing an apartment consisting of a kitchen that measures 10 feet by 6 feet, a room that measures 12 feet by 12 feet, and a room that measures 11 feet by 7 feet. This apartment
    (A) does not comply with the requirements listed in the passage
    (B) complies with the requirements listed in the passage provided that it is not located in the cellar or basement
    (C) complies with the requirements listed in the passage provided that the height of the smaller rooms is at least 9 feet
    (D) may or may not comply with the requirements listed in the passage depending on the clear floor area of the kitchen

53. The definition of the term "living room" that is most in accord with its meaning in the passage is
    (A) a sitting room or parlor
    (B) the largest room in an apartment
    (C) a room used for living purposes
    (D) any room in an apartment containing 120 square feet of clear floor area

54. Assume that one room in a four-room apartment measures 20 feet by 10 feet and contains a dining bay of 8 feet by 6 feet. According to the passage, the dining bay must be provided with a window measuring at least
    (A) 6 square feet
    (B) 7 square feet
    (C) 25 square feet
    (D) 55 square feet

55. Kitchens, according to the preceding passage,
    (A) are not considered "living rooms"
    (B) are considered "living rooms" and must therefore meet the height and area requirements of the passage
    (C) are considered "living rooms" but need meet only the height or area requirements of the passage, not both
    (D) are considered "living rooms" but need not meet area requirements

DIRECTIONS: *The following five questions, numbered 56 through 60, are based on the following regulations.*

EMPLOYEE LEAVE REGULATIONS

"As a full-time, permanent city employee under the Career and Salary Plan, firefighter Peter Smith earns an 'annual leave allowance.' This consists of a certain number of days off a year with pay and can be used for vacation, personal business, or observing religious holidays. As a newly appointed employee, during his first eight years of city service, he will earn an 'annual leave allowance' of 20 days off a year (an average of $1^2/_3$ days off a month). After he has finished eight full years of working for the city, he will begin earning an additional five days off a year. His 'annual leave allowance' will then be 25 days a year and will remain at this amount for seven full years. He will begin earning an additional two days off a year after he has completed a total of 15 years of city employment. Therefore, in his sixteenth year of working for the city, Smith will be earning 27 days off a year as his 'annual leave allowance' (an average of $2^1/_4$ days off a month).

A "sick leave allowance" of one day a month also is given to firefighter Smith, but it can be used only in the case of actual illness. When Smith returns to work after using 'sick leave allowance,' he *must* have a doctor's note if the absence is for a total of more than three days, but he also might be required to show a doctor's note for absences of one, two, or three days."

56. According to the preceding passage, Mr. Smith's "annual leave allowance" consists of a certain number of days off each year that he
    (A) does not get paid for
    (B) gets paid for at time and a half
    (C) can use for personal business
    (D) cannot use for observing religious holidays

57. According to the preceding passage, after Mr. Smith has been working for the city for nine years, his "annual leave allowance" will be
    (A) 20 days a year
    (B) 25 days a year
    (C) 27 days a year
    (D) 37 days a year

58. According to the preceding passage, Mr. Smith will begin earning an average of $2^1/_4$ days off a month as his "annual leave allowance" after he has worked for the city for
    (A) 7 full years
    (B) 8 full years
    (C) 15 full years
    (D) 17 full years

59. According to the preceding passage, Mr. Smith is given a "sick leave allowance" of
    (A) 1 day every 2 months
    (B) 1 day per month
    (C) $1^2/_3$ days per month
    (D) $2^1/_4$ days a month

60. According to the preceding passage, when he uses "sick leave allowance," Mr. Smith might be required to show a doctor's note
  (A) even if his absence is for only one day
  (B) only if his absence is for more than two days
  (C) only if his absence is for more than three days
  (D) only if his absence is for three days or more

DIRECTIONS: *Use the following list of warnings concerning electricity to answer questions 61 through 64.*

- Firefighters never should attempt to cut any wires. They should wait until trained utility workers are available to do any necessary cutting.
- When downed electrical wires are encountered, a danger zone should be established around the downed wire to ensure the safety of firefighters.
- All wires should be treated as though they are charged with a high voltage electrical current.
- Firefighters are in danger of not only shock and burns from electrical equipment but also eye injuries from electrical flashing. A firefighter should never look directly at flashing electrical lines.
- Specific attention must be paid to electrical dangers when raising or lowering hoselines, ladders, or equipment near overhead lines.
- Firefighters should be aware of any tingling sensation felt in the feet when working in an area where wires are down. This sensation might indicate that the ground is charged.
- Full protective clothing should always be worn in areas where electrical hazards might exist. Only approved insulated tools that are regularly tested should be used.

61. When a firefighter is a safe distance away from a downed wire, but the wire is sparking and flashing as the current jumps from the wire to a nearby car, the firefighter should
  (A) cut the wire near the power source
  (B) assume the wire is not charged
  (C) watch the flashing closely for flames
  (D) ignore any tingling felt in the feet
  (E) avoid looking directly at the flashing

62. When in an area in which an electrical hazard might exist, firefighters should wear
  (A) street clothes
  (B) full protective clothing
  (C) only protective gloves
  (D) only protective boots
  (E) only protective gloves and boots

63. Raising a ladder near an overhead power line
  (A) should never be done when there are possible electrical dangers
  (B) is a job that should be conducted only by trained utility workers
  (C) is not a serious concern if a safety zone has been established
  (D) should be done with specific attention to electrical dangers
  (E) is not a concern if a firefighter assumes the wires are not charged

64. When working around a downed wire, a firefighter notices a tingling sensation in her feet. This sensation might indicate
   (A) that the ground is charged
   (B) that the wire is not charged
   (C) that the wire should be cut
   (D) that there is no electrical danger
   (E) that a danger zone is unnecessary

DIRECTIONS: *The following four questions, numbered 65 through 68, are based on the following paragraph.*

> Whenever a social group has become so efficiently organized that it has gained access to an adequate supply of food and has learned to distribute it among its members so well that wealth considerably exceeds immediate demands, it can be depended on to utilize its surplus energy in an attempt to enlarge the sphere in which it is active. The structure of ant colonies renders them particularly prone to this sort of expansionist policy. With very few exceptions, ants of any given colony are hostile to those of any other colony, even of the same species. This condition is bound to produce preliminary bickering among colonies that are closely associated.

65. According to the paragraph, a social group is wealthy when it
   (A) is efficiently organized
   (B) controls large territories
   (C) contains energetic members
   (D) produces and distributes food reserves

66. According to the paragraph, the structure of an ant colony is its
   (A) social organization
   (B) nest arrangement
   (C) territorial extent
   (D) food-gathering activities

67. It follows from the preceding paragraph that the *least* expansionist society would be one that has
   (A) great poverty generally
   (B) more than sufficient wealth to meet its immediate demands
   (C) great wealth generally
   (D) wide inequality between its richest and poorest members

68. According to the preceding paragraph, an ant generally is hostile *except* to other
   (A) insects
   (B) ants
   (C) ants of the same species
   (D) ants of the same colony

DIRECTIONS: Use the following information and table to answer the questions following scenarios A, B, and C (questions 69 through 72).

Carbon monoxide (CO) is a lethal gas that causes more deaths than any other product of combustion. Very small concentrations of carbon monoxide readily combine with chemicals in the blood and quickly cause damage to the brain and other body tissues by blocking the blood's ability to carry oxygen.

TOXIC EFFECTS OF CARBON MONOXIDE

Carbon Monoxide in Air (%)	Symptoms
1.23	Instant unconsciousness; death likely in 1 to 3 minutes
.32	Dizziness, headache, nausea after 5 to 10 minutes; unconsciousness after 30 minutes
.04	Headache after 1 to 2 hours of exposure
.01	No damage—no symptoms

Scenario A

On a very cold day in January, Anthony starts up his car in the garage. Due to the cold temperatures outside, he decides to stay in the car and to leave the garage door closed until the car warms up. After about 10 minutes, Anthony has a headache and starts to feel dizzy and nauseous.

69. According to the information and table presented, the likely cause of Anthony's symptoms is
    (A) the extremely cold temperature
    (B) the lack of carbon monoxide in his blood
    (C) the lack of oxygen in his blood
    (D) carbon monoxide levels of .04%
    (E) carbon monoxide levels of 1.23%

70. According to the information and table, Anthony could become unconscious if he
    (A) has no symptoms
    (B) has a headache three hours later
    (C) gets any more oxygen
    (D) stays in the car for another 20 minutes
    (E) is not exposed to more carbon monoxide

Scenario B

Two firefighters enter a room that contains a smoldering fire. Against the advice of his partner, one of the firefighters takes off his breathing apparatus to clear some debris from the inside of his mask. He instantly falls to the floor unconscious.

71. The likely reason for the firefighter's unconsciousness is
    (A) dizziness, headache, or nausea
    (B) carbon monoxide levels of .01%
    (C) carbon monoxide levels of .04%
    (D) carbon monoxide levels of .32%
    (E) carbon monoxide levels of 1.23%

    Scenario C

    A fire started in a large apartment building and took several hours to extinguish. Evacuation of the building took over an hour to complete. Some of the last people to leave the building were complaining of headaches.

72. What was the likely concentration of carbon monoxide in the apartment building?
    (A) 1.5%
    (B) 1.23%
    (C) .32%
    (D) .04%
    (E) .01%

73. The function of pinion gear (2) in the hand drill shown in the preceding diagram is to
    (A) increase the speed of the chuck
    (B) keep the speed gear from wobbling
    (C) double the turning force on the chuck
    (D) allow reverse rotation of the speed gear

74. Assume that two identical insulated jugs are filled with equal quantities of water from a water tap. A block of ice is placed in one jug and the same quantity of ice, chopped into small cubes, is placed in the other jug. The statement that is most accurate is that the water in the jug containing the chopped ice, compared to the water in the other jug, will be chilled
    (A) faster but to a substantially higher temperature
    (B) faster and to approximately the same temperature
    (C) slower but to a substantially higher temperature
    (D) slower and to approximately the same temperature

75. The long pole and hook shown in the preceding sketch is called a pike pole. Firefighters sometimes push the point and hook through plaster ceilings and then pull the ceiling down. Of the following, the most likely reason for this practice is to
    (A) let heat and smoke escape from the room
    (B) trace defective electric wiring through the house
    (C) see if hidden fire is burning above the ceiling
    (D) remove combustible material that will provide fuel for the fire

DIRECTIONS: *Use the information in the following passage to answer question 76.*

Firefighters Jones and Smith were recently called to a traumatic fire involving a crashed passenger jet. Jones was assigned to a crew responsible for containing and putting out the fire. Smith was assigned to a crew responsible for rescuing victims from the plane.

Many of the victims Smith attempted to save were small children, and many did not survive. In the weeks following the plane crash, Jones noticed that Smith was very depressed and had a hard time focusing his attention on his duties at the station.

76. Considering the recent events, what is most likely the reason for Smith's behavior?
    (A) job stress
    (B) boredom
    (C) drugs
    (D) alcohol
    (E) laziness

77. As your company arrives at the scene of a fire in a large rooming house, a man is spied on the second-story fire escape, about to let himself down to the ground using a sheet he has torn into strips, knotted, and attached to some object in the building. Two middle-aged women and a young boy are observed descending to the street using the fire escape. Flames and smoke are issuing from the third-floor windows. Of the following, the best characterization of this man's behavior is that it is
    (A) intelligent, because the fire has cut off egress by means of the stairway
    (B) not intelligent, because a firm anchor for the knotted sheet could be achieved by securing it to the fire escape
    (C) intelligent, because the fire might heat the fire escape to the point where it is red hot
    (D) not intelligent, because to descend the fire escape would be a quicker and safer means of escape

78. Suppose the same quantity of water is placed in the cup and bowl pictured above and both are left on a table. The time required for the water to evaporate completely from the cup, as compared to the bowl, would be
    (A) longer
    (B) shorter
    (C) equal
    (D) longer or shorter depending on the temperature and humidity in the room

79. Some tools are known as all-purpose tools because they can be used for a great variety of purposes. Others are called special-purpose tools because they are suitable only for a particular purpose. Generally, an all-purpose tool, as compared to a special tool for the same purpose, is
    (A) cheaper
    (B) less efficient
    (C) safer to use
    (D) simpler to operate

80. During a snowstorm, a passenger car with rear-wheel drive gets stuck in the snow. It is observed that the rear wheels are spinning in the snow, and the front wheels are not turning. The statement that best explains why the car is not moving is that
    (A) moving parts of the motor are frozen or blocked by the ice and snow
    (B) the front wheels are not receiving power because of a defective or malfunctioning transmission
    (C) the rear wheels are not obtaining sufficient traction because of the snow
    (D) the distribution of the power to the front and rear wheels is not balanced

81. When the inside face of a rubber suction cup is pressed against a smooth wall, it usually remains in place because of the
    (A) force of molecular attraction between the rubber and the wall
    (B) pressure of the air on the rubber cup
    (C) suction caused by the elasticity of the rubber
    (D) static electricity generated by the friction between rubber and the wall

82. The self-contained breathing apparatus consists of a tank containing breathable air supplied to the user through flexible tubes. This type of breathing apparatus would be *least* effective in an atmosphere containing
    (A) insufficient oxygen to sustain life
    (B) gases irritating to the skin
    (C) gases that cannot be filtered
    (D) a combination of toxic gases

83. At a fire during very cold weather, a firefighter who was ordered to shut down a hoseline left it partially open so that a small amount of water continued to flow out of the nozzle. Leaving the nozzle partially open in this situation was a
    (A) good practice, mainly because the hoseline can be put back into action more quickly
    (B) poor practice, mainly because the escaping water will form puddles, which will freeze on the street
    (C) good practice, mainly because the water in the hoseline will not freeze
    (D) poor practice, mainly because unnecessary water damage to property will result

84. Some sprinkler systems are supplied with water from a single source. Others are supplied with water from two different sources. Generally, firefighters consider the two-source system preferable to the single-source system because the two-source system
    (A) is cheaper to install
    (B) is less likely to be put out of operation
    (C) is capable of delivering water at higher pressure
    (D) goes into operation faster

85. Of the following statements about electric fuses, the one that is most valid is that they
    (A) should never be replaced by coins
    (B) can be replaced by coins for a short time if there are no fuses available
    (C) can be replaced by coins provided that the electric company is notified
    (D) can be replaced by coins provided that care is taken to avoid overloading the circuit

86. The function of the flat surface machined into the shaft in the preceding diagram is to
    (A) prevent slippage of a pulley positioned at this end on the shaft
    (B) provide a nonskid surface to hold the shaft steady as it is machined
    (C) prevent the shaft from rolling about when it is placed on a flat workbench
    (D) reveal subsurface defects in the shaft

DIRECTIONS: Answer questions 87 and 88 on the basis of the diagram at the top of page 261. A, B, C, and D are four meshed gears forming a gear train. Gear A is the driver. Gears A and D each have twice as many teeth as gear B, and gear C has four times as many teeth as gear B. The diagram is schematic; the teeth go all around each gear.

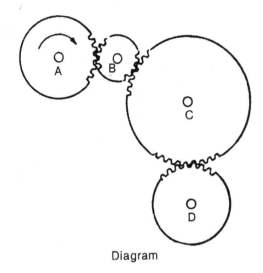

Diagram

87. Two gears that turn in the same direction are
    (A) A and B
    (B) B and C
    (C) C and D
    (D) B and D

88. The two gears that revolve at the same speed are gears
    (A) A and C
    (B) A and D
    (C) B and C
    (D) B and D

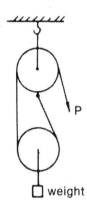

89. In the preceding sketch of a block and fall, if the end of the rope P is pulled so that it moves one foot, the distance the weight will be raised is
    (A) $1/2$ foot
    (B) 1 foot
    (C) $1 1/2$ feet
    (D) 2 feet

1   2   3   4

90. The preceding sketches show four objects that have the same weight but different shapes. The object that is most difficult to tip over is numbered
   (A) 1
   (B) 2
   (C) 3
   (D) 4

91. A firefighter, while taking some clothing to a dry cleaner in his neighborhood, noticed that inflammable cleaning fluid was stored in a way that created a fire hazard. The firefighter called this to the attention of the proprietor, explaining the danger involved. This method of handling the situation was
   (A) bad; the firefighter should not have interfered in a matter that was not his responsibility
   (B) good; the proprietor probably will remove the hazard and be more careful in the future
   (C) bad; the firefighter should have reported the situation to the fire inspector's office without saying anything to the proprietor
   (D) good; because the firefighter was a customer, he should treat the proprietor more leniently than he would treat other violators

92. During the course of fighting a fire, a firefighter encounters an unconscious middle-aged man who is bleeding profusely. It would be most proper for the firefighter to administer first aid by initially
   (A) attempting to bring him back to consciousness
   (B) calling for medical assistance and awaiting its arrival
   (C) administering artificial respiration
   (D) calling for medical assistance and seeking out and blocking the source of the bleeding

93. A firefighter who encounters a victim of shock should give first aid to the victim by
   (A) covering him with a wet cloth to keep him cool
   (B) administering artificial respiration
   (C) loosening his clothing and removing his shoes and stockings
   (D) keeping the patient lying down and elevating extremities if there is no further injury

94. Modern acceptable first-aid technique recommends that, whenever possible, bleeding caused by punctures should be controlled by the application of a
   (A) tourniquet
   (B) pressure bandage
   (C) coagulant
   (D) figure eight bandage

95. A newly appointed firefighter becomes involved in an accident while on duty. Faced with this problem, it would be most appropriate for the firefighter's superior officer to first ascertain
    (A) if the firefighter is accident-prone
    (B) the cause of the accident
    (C) if the orientation program stresses employee safety sufficiently
    (D) if the firefighter has ever been involved in an accident previously that was covered by workmen's compensation

96. A veteran firefighter feels lightheaded soon after reporting for a tour of duty. It would be most proper for the firefighter to
    (A) ask to be excused from ascending heights during this tour of duty
    (B) say nothing because a replacement is not available
    (C) promptly report this condition to his or her superior officer
    (D) discuss the matter with fellow firefighters to see if they have ever had the same problem

97. A firefighter, after completing a special course of study on safety at work, is not convinced that a method taught during the course is the best way to avoid accidents. Upon returning to duty, the firefighter should
    (A) perform the task as it was taught
    (B) perform the task in the manner he or she is convinced is the best way
    (C) attempt to gain the support of fellow firefighters in his or her method
    (D) request a transfer to other duty where it will not be necessary to perform this task

98. A firefighter on inspection duty is offered a sum of money by a store owner to overlook a violation of the fire law. The best action to take would be to
    (A) order the store owner to clear the violation without delay and report the incident to his or her superior officer
    (B) say nothing to the store owner but return to the firehouse to seek the advice of fellow firefighters
    (C) inform the store owner that firefighters do not accept bribes
    (D) refuse the money but offer to assist the store owner to clear the violation during his or her off-duty hours

99. All newly appointed firefighters are put through an extensive training course before they are permitted to participate in the actual activity of fighting fires. The most probable reason for this is that
    (A) it would be unwise for inept firefighters to be exposed to the scrutiny of the public
    (B) the skills necessary for effective firefighting are unique to the job
    (C) the orientation training session is necessary to establish the officer-firefighter relationship
    (D) newly appointed firefighters must be convinced of the dangers inherent in firefighting

100. The statement "Modern technology has greatly influenced firefighting techniques" means most nearly that
    (A) newly appointed firefighters are better equipped to fight fires than their predecessors because they are better educated
    (B) the need for investigating whether changes should be made in procedures becomes important when new firefighting equipment is acquired
    (C) established successful firefighting techniques should be retained whenever possible
    (D) newly acquired equipment always calls for changes in established procedures

Answer Key for Practice Examination 3

1. C	26. B	51. C	76. A
2. D	27. C	52. C	77. D
3. B	28. B	53. C	78. A
4. D	29. D	54. A	79. B
5. C	30. C	55. D	80. C
6. B	31. D	56. C	81. B
7. D	32. A	57. B	82. B
8. D	33. C	58. C	83. C
9. A	34. D	59. B	84. B
10. A	35. B	60. A	85. A
11. D	36. A	61. E	86. A
12. C	37. C	62. B	87. D
13. C	38. A	63. D	88. B
14. A	39. C	64. A	89. A
15. B	40. C	65. D	90. A
16. B	41. C	66. A	91. B
17. C	42. B	67. A	92. D
18. A	43. A	68. D	93. D
19. B	44. B	69. C	94. B
20. D	45. C	70. D	95. B
21. E	46. C	71. E	96. C
22. A	47. A	72. D	97. A
23. E	48. B	73. B	98. A
24. B	49. B	74. B	99. B
25. E	50. C	75. C	100. D

Explanatory Answers for Practice Examination 3

1. **(C)** This is stated in the first sentence of the first paragraph.

2. **(D)** This is stated in the first sentence of the third paragraph.

3. **(B)** This is stated in the second sentence of the fifth paragraph.

4. **(D)** 75 companies × 2 hours each = 150 hours × 6 persons per engine company = 900 personnel hours.

5. **(C)** See the last paragraph.

6. **(B)** If force is necessary to move people away from hydrants to shut the hydrants, the police department should be notified for assistance. The firefighter should not become involved in this action.

7. **(D)** To effectively fight a fire, firefighters should be in good physical condition. One of the lieutenant's responsibilities is to keep the firefighters from becoming injured.

8. **(D)** "A potentially explosive situation developed rapidly." There was no reason for the firefighters to hang around waiting for trouble.

9. **(A)** 75 companies × 2 hours = 150 hours.

10. **(A)** The police were called by the lieutenant in command of Engine 299 to control the crowd and to protect firefighters, not to close the hydrant.

11. **(D)** 1,100 gallons per minute × 120 minutes (2 hours) = 132,000 gallons.

12. **(C)** See the beginning of the fourth paragraph.

13. **(C)** 600 hydrants ÷ 75 companies = 8 hydrants.

14. **(A)** This is stated in the last sentence of the fourth paragraph.

15. **(B)** This is stated in the second sentence of the fifth paragraph.

16. **(B)** For every one gallon of expansion concentrate, 50 to 300 gallons of foam are produced. Thus, if three gallons of concentrate are used, 150 (50 × 3) to 900 (300 × 3) gallons of foam will be produced.

17. **(C)** The total damaged area of the building that was damaged is the sum of the square feet consumed on the three floors. Thus, 1,300 + 12,000 + 7,500 = 20,800 square feet.

18. **(A)** 1,000 gallons is 1/10 of 10,000 gallons. If 10,000 gallons are pumped per hour, we want to know what 1/10 of one hour is. Because there are 60 minutes in 1 hour, 60 × 1/10 = 6 minutes.

19. **(B)** 900/75 = 12. To check your answer, 75 × 12 = 900.

20. **(D)** If it takes a 24-foot ladder to reach the second story, it takes 12 feet of ladder for each story (24/2 = 12). Because there are five stories, 12 × 5 = 60.

21. **(E)** Because there are two hydrants for every block and 75 blocks for every square mile, there are 75 × 2 = 150 hydrants for every square mile. Because we are looking at 2 square miles, 150 × 2 = 300 hydrants for 2 square miles.

22. **(A)** The correct spelling is "numerous."

23. **(E)** The correct spelling is "synthetic."

24. **(B)** The word "is" is the best word to complete the sentence.

25. **(E)** The word "begun" is the best word to complete the sentence. This is the correct past-tense form of the word "begin."

26. **(B)** Generally, there are many new students in a school at the beginning of the school year who are unfamiliar with the exits. By having the majority of drills in the early part of the school year, the children become familiar with the evacuation plan and the location of the exits.

27. **(C)** Toxic gases and smoke from a fire are light and rise rapidly through vertical openings in a building. They "mushroom" on the upper floors. This is why ventilation at the roof is essential.

28. **(B)** People frequently create unsafe situations. Poor housekeeping or poor safety habits eventually could result in a destructive fire. Educating people about the existing dangers can save them from injury, death, and destruction of their property.

29. **(D)** A top priority in firefighting is ventilation of the building to remove smoke and toxic gases. Smoke removal heightens visibility, saves lives, reduces the chances of panic, and enables firefighters to perform more effectively.

30. **(C)** Only authorized, qualified firefighters are permitted to operate at a fire. Be firm. If the man insists upon "helping," have a police officer remove him from the scene.

31. **(D)** Generally, it is a good policy to obtain as much information as you can from the complainant. It makes reporting easier and more complete. By having the full name and address of the complainant of record, contact could be made quite easily if necessary.

32. **(A)** When returning from an alarm, fire apparatus should adhere to traffic regulations just like any other vehicle on the road. Sirens can cause motorists and pedestrians to become confused and nervous, thereby increasing the chance of an accident.

33. **(C)** People using a fire escape in an emergency usually are excited. Any obstruction in their path can present a hazard. By having the platform level with the sill, people using the fire escape have better footing, reducing the chance of tripping.

34. **(D)** The alarm ringing and the discharge of water from the pipe are indications that one or more sprinkler heads have opened and there is water flow in the building. Because the location of the firehouse is known and nearby and the location of the alarm box is uncertain, the wise decision is to go to the firehouse and report the situation.

35. **(B)** If a hoseline is run between the rungs of a ladder, the ladder no longer is maneuverable until the hoseline is shut down and disconnected.

36. **(A)** Notifying the police department is the quickest and most proper action to take. Radio communications enable the police to be at the scene rapidly. They are trained to handle this type of emergency. You must remain available to respond to fires.

37. **(C)** The immediate administration of first aid by the firefighter to restore breathing gives the victim a chance of surviving. The firefighter also should direct a passerby to call for medical assistance, either by telephone or at the change booth. If a transit officer is nearby, notify the officer.

38. **(A)** By conducting a drill or practice at the site of a possible emergency, firefighting forces become familiar with the various aspects of a particular type of building or location. They get firsthand knowledge of the obstacles that might be encountered during an actual emergency.

39. **(C)** Lumberyards store a wide assortment of wood and wood products. Strong upward drafts can develop at a lumberyard fire and can carry sparks and large brands a considerable distance, especially if there is much wind.

40. **(C)** At a suspicious fire, the officer in charge should document pertinent facts as soon as possible. The passing of time generally results in a witness forgetting many of the details of what he or she saw originally. Documenting the facts will enable the firefighter to recall the incident.

41. **(C)** By asking the theater manager to correct the situation, the firefighter is making the manager aware of the existing violation. It is the manager's responsibility to correct the situation.

42. **(B)** An uncharged 50-foot length of $2^1/_2$-inch hose weighs approximately 40 pounds. Water weighs $8^1/_3$ pounds per gallon, and there are 13 gallons in the 50-foot length. Obviously, the uncharged line is much lighter and much easier to carry to the point where water is needed.

43. **(A)** Smoke and hot gases from the fire will rise and "mushroom" at the roof or upper floors. If the roof is not vented, the hot gases will ignite and spread fire to the roof area.

44. **(B)** The fire equipment must be maintained to respond immediately in an emergency. Delays caused by apparatus breakdowns could result in loss of life or excessive property damage. This is why extensive preventative maintenance is performed on high-quality department apparatus.

45. **(C)** Smoke is a product of combustion, so eliminate (D). The newspaper carries reports of death from smoke inhalation, so eliminate (A). Smoke, as the product of combustion, not only reduces the oxygen content of the air but contains particles, carbon monoxide, and whatever toxic gases might be produced by the substance burning.

46. **(C)** As long as the firefighter is performing his or her job properly, there is no reason to be intimidated by the owner.

47. **(A)** Charging a property owner a fee for calling the fire department is not a good policy. People might be unable or unwilling to afford the fee and might attempt to extinguish the fire themselves. Experience has shown that delays in calling the fire department have resulted in fires getting out of control, causing loss of life and extensive property damage.

48. **(B)** The use of dry pipe standpipe systems where the pipes are exposed to freezing temperatures avoids having pipes blocked by ice or pipes that burst from the pressure of frozen water.

49. **(B)** Calling in the alarm will bring firefighters to the reported location much faster than if the firefighter ran to the house to see if there really was a fire. Every report of fire must be taken seriously.

50. **(C)** The passage indicates that three tugs stands for T, which indicates that the individual wants to be taken up.

51. **(C)** A room must have at least 70 square feet of floor space, but if it has less than 80 square feet, it must be at least 9 feet high. 70 square feet × 9 feet = 630 cubic feet.

52. **(C)** This is stated in paragraph b. Every living room with less than 80 square feet of floor area shall be 9 feet high.

53. **(C)** When speaking of all rooms in apartments of three or more rooms as "living rooms," the passage must be referring to rooms used for living.

54. **(A)** The floor space of the dining bay is 6 × 8 = 48 square feet. One-eighth of 48 = 6 square feet.

55. **(D)** The words "every other living room except a kitchen" imply that a kitchen is a living room but that it is not subject to the size requirements of other living rooms.

56. **(C)** See the second sentence.

57. **(B)** For 8 to 15 years of service, the annual leave is 25 days.

58. **(C)** The sixteenth year, in which leave is earned at the rate of $2\frac{1}{4}$ days per month, comes after 15 full years.

59. **(B)** See the first sentence of the second paragraph.

60. **(A)** See the last sentence.

61. **(E)** The fourth warning on the list states that firefighters should never look directly at flashing electrical lines.

62. **(B)** The last warning on the list states that full protective clothing always should be worn in areas in which electrical hazards might exist.

63. **(D)** The fifth warning on the list states that specific attention must be paid to electrical dangers when raising a ladder near an overhead power line.

64. **(A)** The sixth warning on the list states that any tingling sensation felt in the feet might indicate that the ground is charged.

65. **(D)** When supply exceeds current needs, the society is wealthy and can distribute its surplus.

66. **(A)** This information can be inferred from the first and second sentences.

67. **(A)** Poverty is the opposite of wealth. If there is poverty, there is no surplus and no expansionism.

68. **(D)** This is stated in the third sentence.

69. **(C)** The table lists headache, dizziness, and nausea as signs that someone has been exposed to an atmosphere with carbon monoxide levels of .32% for 5 to 10 minutes. The paragraph indicates that carbon monoxide causes damage by blocking the blood's ability to carry oxygen. Thus, (C) is the correct response.

70. **(D)** The table lists unconsciousness as a symptom of remaining in an atmosphere with carbon monoxide levels of .32% for more than 30 minutes. Because Tony already exhibits the symptoms of headache, dizziness, and nausea after 10 minutes, he could become unconscious if he remains in the car for an additional 20 minutes.

71. **(E)** The table lists instant unconsciousness as a symptom of breathing air with carbon monoxide levels of 1.23%. Thus, (E) is the correct response.

72. **(D)** The table lists headache as a symptom of exposure to carbon monoxide levels of .04% for 1 to 2 hours. Because the last people to leave the building were in it for over an hour and were complaining of headaches, (D) is the correct response.

73. **(B)** The pinion gear is a small bevel or spur gear that meshes with the speed gear. The pinions "bind" the large gear to prevent it from wobbling.

74. **(B)** The small cubes of chopped ice have a greater surface area than the large block of ice. Therefore, more water will be exposed to the area of the chopped ice, and it will chill faster to the same temperature.

75. **(C)** In a fire, flame and hot gases can enter cracks or holes in a ceiling or wall and can burn undetected. If the ceilings are not opened and the fire extinguished, the fire will continue to burn and spread throughout the structure. The pike pole is used for this purpose.

76. **(A)** The passage indicates that the plane crash was very traumatic. Smith was assigned to rescue victims from the plane, many of whom did not survive. Judging from the information presented in the passage, it is likely that Smith's depression and lack of focus are due to job stress.

77. **(D)** There are people on the fire escape right now, so it obviously is not too hot. The fire escape would be quicker and safer than a sheet, which might pull loose, tear, or burn. During moments of panic, people sometimes use poor judgment.

78. **(A)** Evaporation occurs when the molecules near the surface of the liquid escape into the air. The water in the bowl has a greater surface area exposed to the air, causing it to evaporate faster than the water in the cup.

79. **(B)** The special-purpose tool generally is more efficient because it is designed for one specific purpose. The all-purpose tool can accomplish the intended job, but it usually requires more time and works less effectively. In firefighting, time and efficiency are essential.

80. **(C)** For the rear wheels (which are the driving wheels) to drive or push the car, friction or resistance must be present between the tires and the road surface. The ice and snow create a slippery road condition, causing the tires to lose traction. The front wheels do not turn because power from the engine is transmitted to the rear wheels only.

81. **(B)** When the suction cup is pressed against the wall, the air in the cup is forced out and a vacuum or negative pressure remains. Atmospheric air with a pressure of 14.7 psi (pounds per square inch) pushes against the rubber cup and holds it against the wall.

82. **(B)** Many toxic gases and acids are absorbed directly through the skin. To operate safely and effectively in this type of atmosphere, self-contained suits made of rubber or other nonporous material should be used in conjunction with the self-contained breathing apparatus. The self-contained apparatus takes in no outside air, so filtration is not a problem.

83. **(C)** During very cold weather, water will freeze much more quickly when it is idle than when it is moving. By keeping the water flowing, there is a smaller chance that the water in the hoseline will freeze.

84. **(B)** Some sprinkler systems require two sources of water. In the event that one source is placed out of service, the other is available to keep the system operating. Two-source systems can be supplied by: a) two water mains, each on a different street; b) a street main and a gravity tank; c) a street main and a pressure tank.

85. **(A)** The function of a fuse is to protect against overload or overheating of a circuit. A burned-out fuse indicates that a dangerous condition on a circuit has been avoided. By replacing the fuse with a coin that cannot burn out, you bypass the safety feature of the fuse and create a fire hazard.

86. **(A)** The flat surface of the shaft is called a keyway. A corresponding opening in a gear or pulley accommodates the keyway. This prevents slippage of the gear or pulley on the shaft.

87. **(D)** Adjacent gears turn in opposite directions. Gear A turns in a clockwise direction; therefore, Gear B, which is adjacent to it, will turn in a counterclockwise direction. Continuing, Gear C, which is adjacent to Gear B, will turn in a clockwise direction; Gear D, which is adjacent to Gear C, will turn counterclockwise, the same as Gear B.

88. **(B)** Gears of the same size connected into the same system will turn at the same speed.

89. **(A)** The sketch is of a fixed pulley and a moveable pulley. With this arrangement, the effort to move the weight will be one-half. The effort distance (the length of the pull), however, will be twice the distance the weight travels. Therefore, a one-foot pull raised the weight one-half foot.

90. **(A)** With all the objects equal in weight, the pyramid is the most difficult to tip over. This is due to the fact that the bulk of the volume and weight is at the base.

91. **(B)** Surely the proprietor of the dry cleaning store does not want to create a dangerous situation for himself and for his property. He will be grateful for the information and will most likely change his manner of storing cleaning fluid. The firefighter's action falls into the category of friendly, helpful neighbor.

92. **(D)** Profuse bleeding can be life threatening. While waiting for medical assistance, the firefighter should initially attempt to control the bleeding by applying direct pressure to the wound with a sterile dressing if one is available. If bleeding persists, the firefighter should reach into the wound and pinch the blood vessel.

93. **(D)** Proper treatment for a person in shock includes ensuring adequate breathing, controlling the bleeding, administering oxygen, elevating extremities, avoiding rough handling, preventing loss of body heat, keeping the victim lying down, and giving nothing by mouth.

94. **(B)** A pressure bandage will be sufficient. Generally, when a person suffers a puncture wound, there is very little external bleeding. However, severe internal bleeding is possible.

95. **(B)** Before any corrective action can be taken, the cause of the accident must be known. By getting the facts, the officer is able to determine why the accident occurred. Knowing why the accident occurred can prevent a similar accident in the future.

96. **(C)** The superior officer, being aware of the firefighter's condition, can summon medical advice for the firefighter and, if necessary, can relieve him or her from duty. Getting lightheaded or dizzy in a hazardous situation can result in serious injury to the firefighter and to anyone he or she might be assisting.

97. **(A)** The information presented at the special course on safety apparently was compiled by using case studies, information gathered from accident reports, and firsthand experience. The firefighter could voice his or her opinion on the subject to the safety instructor or to his or her superior officers. However, the firefighter should perform the task as it was taught.

98. **(A)** A firefighter on inspection duty has a job to perform. That job is to note and report violations and to follow through to be certain violations are corrected. Accepting a bribe to overlook violations is a criminal act as well as a dereliction of duty.

99. **(B)** An extensive training program is necessary because of the nature of the occupation of firefighting. Fires and fire conditions vary with each situation. The tremendous hazards present when fighting fires are not found in any other occupation. The lives of the fire victims as well as those of fellow firefighters are at stake. For the firefighter to function effectively, he or she must be as well-trained as possible before being permitted to take part in actual firefighting.

100. **(D)** Changes in procedures are necessary due to the many technological advances that have been introduced into the field of firefighting. These technological advances enable the firefighter to function more efficiently and with greater effectiveness. Some advances that have dictated changes include improved self-contained breathing apparatus, handi-talkies, power saws, hydraulic cutting equipment, and sophisticated foams and fog nozzles.

PART THREE

CONTENTS

Practice Examination 4   277
Answers and Explanations   296
Practice Examination 5   307
Answers and Explanations   347
Appendix   363

Answer Sheet for Practice Examination 4

1. Ⓐ Ⓑ Ⓒ Ⓓ Ⓔ
2. Ⓐ Ⓑ Ⓒ Ⓓ Ⓔ
3. Ⓐ Ⓑ Ⓒ Ⓓ Ⓔ
4. Ⓐ Ⓑ Ⓒ Ⓓ Ⓔ
5. Ⓐ Ⓑ Ⓒ Ⓓ Ⓔ
6. Ⓐ Ⓑ Ⓒ Ⓓ Ⓔ
7. Ⓐ Ⓑ Ⓒ Ⓓ Ⓔ
8. Ⓐ Ⓑ Ⓒ Ⓓ Ⓔ
9. Ⓐ Ⓑ Ⓒ Ⓓ Ⓔ
10. Ⓐ Ⓑ Ⓒ Ⓓ Ⓔ
11. Ⓐ Ⓑ Ⓒ Ⓓ Ⓔ
12. Ⓐ Ⓑ Ⓒ Ⓓ Ⓔ
13. Ⓐ Ⓑ Ⓒ Ⓓ Ⓔ
14. Ⓐ Ⓑ Ⓒ Ⓓ Ⓔ
15. Ⓐ Ⓑ Ⓒ Ⓓ Ⓔ
16. Ⓐ Ⓑ Ⓒ Ⓓ Ⓔ
17. Ⓐ Ⓑ Ⓒ Ⓓ Ⓔ
18. Ⓐ Ⓑ Ⓒ Ⓓ Ⓔ
19. Ⓐ Ⓑ Ⓒ Ⓓ Ⓔ
20. Ⓐ Ⓑ Ⓒ Ⓓ Ⓔ
21. Ⓐ Ⓑ Ⓒ Ⓓ Ⓔ
22. Ⓐ Ⓑ Ⓒ Ⓓ Ⓔ
23. Ⓐ Ⓑ Ⓒ Ⓓ Ⓔ
24. Ⓐ Ⓑ Ⓒ Ⓓ Ⓔ
25. Ⓐ Ⓑ Ⓒ Ⓓ Ⓔ
26. Ⓐ Ⓑ Ⓒ Ⓓ Ⓔ
27. Ⓐ Ⓑ Ⓒ Ⓓ Ⓔ
28. Ⓐ Ⓑ Ⓒ Ⓓ Ⓔ
29. Ⓐ Ⓑ Ⓒ Ⓓ Ⓔ
30. Ⓐ Ⓑ Ⓒ Ⓓ Ⓔ
31. Ⓐ Ⓑ Ⓒ Ⓓ Ⓔ
32. Ⓐ Ⓑ Ⓒ Ⓓ Ⓔ
33. Ⓐ Ⓑ Ⓒ Ⓓ Ⓔ
34. Ⓐ Ⓑ Ⓒ Ⓓ Ⓔ
35. Ⓐ Ⓑ Ⓒ Ⓓ Ⓔ
36. Ⓐ Ⓑ Ⓒ Ⓓ Ⓔ
37. Ⓐ Ⓑ Ⓒ Ⓓ Ⓔ
38. Ⓐ Ⓑ Ⓒ Ⓓ Ⓔ
39. Ⓐ Ⓑ Ⓒ Ⓓ Ⓔ
40. Ⓐ Ⓑ Ⓒ Ⓓ Ⓔ
41. Ⓐ Ⓑ Ⓒ Ⓓ Ⓔ
42. Ⓐ Ⓑ Ⓒ Ⓓ Ⓔ
43. Ⓐ Ⓑ Ⓒ Ⓓ Ⓔ
44. Ⓐ Ⓑ Ⓒ Ⓓ Ⓔ
45. Ⓐ Ⓑ Ⓒ Ⓓ Ⓔ
46. Ⓐ Ⓑ Ⓒ Ⓓ Ⓔ
47. Ⓐ Ⓑ Ⓒ Ⓓ Ⓔ
48. Ⓐ Ⓑ Ⓒ Ⓓ Ⓔ
49. Ⓐ Ⓑ Ⓒ Ⓓ Ⓔ
50. Ⓐ Ⓑ Ⓒ Ⓓ Ⓔ
51. Ⓐ Ⓑ Ⓒ Ⓓ Ⓔ
52. Ⓐ Ⓑ Ⓒ Ⓓ Ⓔ
53. Ⓐ Ⓑ Ⓒ Ⓓ Ⓔ
54. Ⓐ Ⓑ Ⓒ Ⓓ Ⓔ
55. Ⓐ Ⓑ Ⓒ Ⓓ Ⓔ
56. Ⓐ Ⓑ Ⓒ Ⓓ Ⓔ
57. Ⓐ Ⓑ Ⓒ Ⓓ Ⓔ
58. Ⓐ Ⓑ Ⓒ Ⓓ Ⓔ
59. Ⓐ Ⓑ Ⓒ Ⓓ Ⓔ
60. Ⓐ Ⓑ Ⓒ Ⓓ Ⓔ
61. Ⓐ Ⓑ Ⓒ Ⓓ Ⓔ
62. Ⓐ Ⓑ Ⓒ Ⓓ Ⓔ
63. Ⓐ Ⓑ Ⓒ Ⓓ Ⓔ
64. Ⓐ Ⓑ Ⓒ Ⓓ Ⓔ
65. Ⓐ Ⓑ Ⓒ Ⓓ Ⓔ
66. Ⓐ Ⓑ Ⓒ Ⓓ Ⓔ
67. Ⓐ Ⓑ Ⓒ Ⓓ Ⓔ
68. Ⓐ Ⓑ Ⓒ Ⓓ Ⓔ
69. Ⓐ Ⓑ Ⓒ Ⓓ Ⓔ
70. Ⓐ Ⓑ Ⓒ Ⓓ Ⓔ
71. Ⓐ Ⓑ Ⓒ Ⓓ Ⓔ
72. Ⓐ Ⓑ Ⓒ Ⓓ Ⓔ
73. Ⓐ Ⓑ Ⓒ Ⓓ Ⓔ
74. Ⓐ Ⓑ Ⓒ Ⓓ Ⓔ
75. Ⓐ Ⓑ Ⓒ Ⓓ Ⓔ
76. Ⓐ Ⓑ Ⓒ Ⓓ Ⓔ
77. Ⓐ Ⓑ Ⓒ Ⓓ Ⓔ
78. Ⓐ Ⓑ Ⓒ Ⓓ Ⓔ
79. Ⓐ Ⓑ Ⓒ Ⓓ Ⓔ
80. Ⓐ Ⓑ Ⓒ Ⓓ Ⓔ
81. Ⓐ Ⓑ Ⓒ Ⓓ Ⓔ
82. Ⓐ Ⓑ Ⓒ Ⓓ Ⓔ
83. Ⓐ Ⓑ Ⓒ Ⓓ Ⓔ
84. Ⓐ Ⓑ Ⓒ Ⓓ Ⓔ
85. Ⓐ Ⓑ Ⓒ Ⓓ Ⓔ
86. Ⓐ Ⓑ Ⓒ Ⓓ Ⓔ
87. Ⓐ Ⓑ Ⓒ Ⓓ Ⓔ
88. Ⓐ Ⓑ Ⓒ Ⓓ Ⓔ
89. Ⓐ Ⓑ Ⓒ Ⓓ Ⓔ
90. Ⓐ Ⓑ Ⓒ Ⓓ Ⓔ
91. Ⓐ Ⓑ Ⓒ Ⓓ Ⓔ
92. Ⓐ Ⓑ Ⓒ Ⓓ Ⓔ
93. Ⓐ Ⓑ Ⓒ Ⓓ Ⓔ
94. Ⓐ Ⓑ Ⓒ Ⓓ Ⓔ
95. Ⓐ Ⓑ Ⓒ Ⓓ Ⓔ
96. Ⓐ Ⓑ Ⓒ Ⓓ Ⓔ
97. Ⓐ Ⓑ Ⓒ Ⓓ Ⓔ
98. Ⓐ Ⓑ Ⓒ Ⓓ Ⓔ
99. Ⓐ Ⓑ Ⓒ Ⓓ Ⓔ
100. Ⓐ Ⓑ Ⓒ Ⓓ Ⓔ

TEAR HERE

PRACTICE EXAMINATION 4

The time allowed for the entire examination is 3½ hours.

DIRECTIONS: *Each question has four suggested answers lettered A, B, C, and D. Decide which one is the best answer. On the sample answer sheet, find the question number that corresponds to the answer you have selected and darken the area with a soft pencil.*

1. While performing a routine inspection of a factory building, a firefighter is asked a question by the plant manager about a matter under the control of the health department and about which the firefighter has little knowledge. In this situation, the best course of action for the firefighter to take is to
   (A) answer the question to the best of his or her knowledge
   (B) tell the manager that he or she is not permitted to answer the question because it does not relate to a fire department matter
   (C) tell the manager that it will be referred to the health department
   (D) suggest to the manager that he communicate with the health department about the matter

2. A firefighter on duty who answers a departmental telephone should give his or her name and rank
   (A) at the start of the conversation, as a matter of routine
   (B) only if asked for this information by the caller
   (C) only if the caller is a superior officer
   (D) only if the telephone message requires the firefighter to take some action

3. At a five-alarm fire in the West End, several companies from Northside were temporarily assigned to occupy the quarters and take over the duties of companies engaged in fighting the fire. The main reason for relocating the Northside companies was to
   (A) protect the firehouses from robbery or vandalism that might occur if they were left vacant for a long period of time
   (B) provide for speedy response to the fire if additional companies were required
   (C) give the Northside companies an opportunity to become familiar with the problems of the West End area
   (D) provide protection to the West End area in the event that other fires occurred

4. Two firefighters, while on their way to report for duty early one morning, observe a fire in a building containing a supermarket on the street level and apartments on the upper stories. One firefighter runs into the building to spread the alarm to tenants. The other firefighter runs to a street alarm box two blocks away and sends an alarm. The latter firefighter should then
   (A) return to the building on fire and help evacuate the tenants
   (B) remain at the fire alarm box to direct the first fire company that arrives to the location of the fire
   (C) look for a telephone to call his own fire company and explain that he and his companion will be late in reporting for duty
   (D) look for a telephone to call the health department and request that an inspector be sent to the supermarket to examine the food involved in the fire

5. A man found an official fire department badge and gave it to his young son to use as a toy. The man's action was improper mainly because
   (A) it is disrespectful to the fire department to use the badge in this manner
   (B) the boy might injure himself playing with the badge
   (C) an effort first should have been made to locate the owner of the badge before giving it to the boy
   (D) the badge should have been returned to the fire department

6. In the case of a fire in a mailbox, the fire department recommends that an extinguishing agent that smothers the fire, such as carbon tetrachloride, be used. Of the following, the most likely reason for not recommending the use of water is that
   (A) water is not effective on fires in small, tightly enclosed spaces
   (B) someone might have mailed chemicals that could explode in contact with water
   (C) water might damage the mail untouched by fire so that it could not be delivered
   (D) the smothering agent can be put on the fire faster than water can be

7. Of the following, the main difficulty in obtaining accurate information about the causes of fires is that
   (A) firefighters are too busy putting out fires to have time to investigate their causes
   (B) most people have little knowledge of fire hazards
   (C) fires destroy much of the evidence that would indicate the causes of the fires
   (D) fire departments are more interested in fire prevention than in investigating fires that already have occurred

8. In an effort to discourage the sending of false alarms and to help apprehend people guilty of this practice, it is suggested that the handles of fire alarm boxes be covered with a dye that would stain the hand of a person sending an alarm and would not wash off for 24 hours. The dye would be visible only under an ultraviolet light. Of the following, the chief objection to such a device is that it would
   (A) require funds that can be better used for other purposes
   (B) have no effect on false alarms transmitted by telephone
   (C) discourage some persons from sending alarms for real fires
   (D) punish the innocent as well as the guilty

9. Automatic fire-extinguishing sprinkler systems sometimes are not effective on fires accompanied by explosions chiefly because
   (A) these fires do not generate enough heat to start sprinkler operation
   (B) the pipes supplying the sprinklers usually are damaged by the explosion
   (C) fires in explosive materials usually cannot be extinguished by water
   (D) sprinkler heads usually are clogged by dust created by the explosion

10. When a fire occurs in the vicinity of a subway system, there is the possibility that water from the firefighters' hose streams will flood underground portions of the subway lines through sidewalk gratings. Of the following methods of reducing this danger, the one that generally would be suitable is for the officer in command to order subordinates to
    (A) use fewer hoselines and smaller quantities of water than they would ordinarily
    (B) attack the fire from positions that are distant from the sidewalk gratings
    (C) cover the sidewalk gratings with canvas tarpaulins
    (D) advise the subway dispatcher to reroute the subway trains

11. When responding to alarms, fire department apparatus generally follow routes established in advance. The *least* valid justification for this practice is that
    (A) motorists living in the area become familiar with these routes and tend to avoid them
    (B) the likelihood of collision between two pieces of fire department apparatus is reduced
    (C) the fastest response generally is obtained
    (D) road construction, road blocks, detours, and similar conditions can be avoided

12. From a distance, an off-duty firefighter sees a group of teenage boys set fire to a newspaper and then toss the flaming pages into the open window of a building being torn down. In this situation, the first action that should be taken by the firefighter is to
    (A) send a fire alarm from the closest street alarm box
    (B) chase the boys and attempt to catch one of them
    (C) investigate whether a fire has been started
    (D) call the police from the closest police alarm box or telephone

13. When responding to an alarm, officers are not to talk to the driver except to give orders or directions. Of the following, the best justification for this rule is that it
    (A) gives the officer an opportunity to make preliminary plans for handling the fire problem
    (B) enables the driver to concentrate on driving the apparatus
    (C) maintains the proper relationship between ranks while on duty
    (D) permits the officer to observe the driver's skill, or lack thereof, in driving the apparatus

14. The approved method of reporting a fire by telephone in most major cities is to dial
    (A) the central headquarters of the fire department
    (B) 911
    (C) the local fire station
    (D) the telephone operator

15. If there should be two fire hydrants for every square city block and there are 75 square city blocks in a mile, what percentage of a mile would be covered by 120 fire hydrants?
    (A) 60%
    (B) 70%
    (C) 80%
    (D) 90%
    (E) 100%

16. A fire has started 3.2 miles from the firehouse. How long will it take the firefighters to arrive at the fire if they travel at an average speed of 30 mph?
    (A) 6.6 minutes
    (B) 7.1 minutes
    (C) 8.3 minutes
    (D) 8.5 minutes
    (E) 9.2 minutes

17. $13^2 + (9 - 7) =$
    (A) 178
    (B) 176
    (C) 160
    (D) 162
    (E) 171

18. If water pressure (*P*) = 500, water temperature (*W*) = 45, and spray angle = 3.2, what is the correct answer to the equation *P/A* − 2(*W*) = ?
    (A) 23.4
    (B) 42
    (C) 56.32
    (D) 66.25
    (E) 74.3

19. In Hewson County, Missouri, last year, there were calls for 21 house fires, 17 industrial fires, 2 industrial accidents, and 46 car accidents. Fire station A received 34 of these calls last year. What percentage of the total number of calls did station A receive?
    (A) 23%
    (B) 35%
    (C) 40%
    (D) 42%
    (E) 46%

DIRECTIONS: *Choose the correct spelling of the word to complete the sentences in questions 20 and 21.*

20. Extensive training is necessary for a firefighter to become _____ at using the equipment involved in firefighting.
    (A) profishant
    (B) proficient
    (C) prophicent
    (D) profiscent
    (E) proficent

21. Firefighters might have difficulty locating _____ victims in a smoke-filled building.
    (A) unconshous
    (B) unconscus
    (C) unconscious
    (D) unconshious
    (E) unconscous

DIRECTIONS: *Choose the appropriate word or phrase to complete the sentences in questions 22 through 24.*

22. Unfortunately, many of the clues leading to the cause of the fire were _____ beneath the rubble.
    (A) hidden
    (B) hid
    (C) hide
    (D) hidded
    (E) hides

23. Firefighters Johnson and Fernandez helped many people escape from a large industrial fire. They were later rewarded for _____ heroic efforts.
  (A) they
  (B) them
  (C) they're
  (D) their
  (E) there

24. It often is better for at least two firefighters to attempt to _____ a ladder.
  (A) raise
  (B) rise
  (C) rose
  (D) risen
  (E) raised

25. Suppose you are a firefighter on housewatch duty when a civilian enters the firehouse. He introduces himself as a British firefighter visiting the country to study American firefighting methods. He asks you for permission to ride on the fire apparatus when it responds to alarms so he can observe operations firsthand. You know it is against departmental policy to permit civilians to ride apparatus without written permission from headquarters. In this situation, you should
  (A) refuse the request but suggest that he follow the apparatus in his own car when it responds to an alarm
  (B) call headquarters and request permission to permit the visitor to ride the apparatus
  (C) refuse the request and suggest that he apply to headquarters for permission
  (D) refuse the request and suggest that he return the next time the fire department holds an open house

26. When operating at a pier fire, firefighters usually avoid driving their apparatus onto the pier itself. The main reason for this precaution is to reduce the possibility that the apparatus will be
  (A) delayed in returning to quarters
  (B) driven off the end of the pier
  (C) destroyed by a fire that spreads rapidly
  (D) in the way of the firefighters

27. Pumpers purchased by the fire department are equipped with enclosed cabs. In the past, fire department apparatus were open with no cab or roof. The main advantage of the enclosed cab is that it provides
  (A) additional storage space for equipment
  (B) a place of shelter for firefighters operating in an area of radioactivity
  (C) protection for firefighters from weather conditions and injury
  (D) emergency first-aid and ambulance facilities

28. Heavy blizzards greatly increase the problems and work of the fire department. When such a situation occurs, the fire commissioner could reasonably be expected to
  (A) order members of the fire department to perform extra duty
  (B) limit parking on city streets
  (C) station firefighters at fire alarm boxes to prevent the sending of false alarms
  (D) prohibit the use of kerosene heaters

29. Regulations of the fire department require that, when placing a hose on a fire wagon, care should be taken to avoid bending the hose at places where it had been bent previously. The most important reason for this requirement is that repeated bending of a hose at the same places will cause
    (A) kinks in the hose at those places
    (B) weakening of the hose at those places
    (C) discoloration of the hose at those places
    (D) dirt to accumulate and clog the hose at those places

30. While fighting a fire in an apartment when the occupants are not at home, a firefighter finds a sum of money in a closet. Under these circumstances, the firefighter should turn over the money to
    (A) a responsible neighbor
    (B) the desk sergeant of the nearest police station
    (C) the superintendent of the apartment house
    (D) his or her superior officer

31. "When it is necessary to remove a cornice, every effort should be made to pull it back on the roof." The most important reason for the direction in this quotation is that pulling back the cornice on the roof rather than dropping it to the street below
    (A) requires less time
    (B) is safer for the people on the street
    (C) makes it possible to reuse the cornice
    (D) is less dangerous to firefighters working on the roof

32. After a fire has been extinguished, one firefighter often remains at the scene after the others have left. Of the following, the main reason for this practice is that this firefighter can
    (A) prevent looters from stealing valuables
    (B) watch for any rekindling of the fire
    (C) search the area for lost valuables
    (D) examine the premises for evidence of arson

33. Firefighters usually attempt to get as close as possible to the seat of a fire so they can direct their hose streams with accuracy. Intense heat, however, sometimes keeps them at a distance. An unsatisfactory method of overcoming this problem is to
    (A) have firefighters use a solid object, such as a wall, as a shield
    (B) keep firefighters cool by wetting them with small streams of water
    (C) use large high-pressure streams and operate at a great distance from the fire
    (D) use a water spray to break down the heat waves coming from the fire

34. Suppose you are an off-duty firefighter driving your car in the midtown area. As you cross an intersection, you hear sirens and, looking back, see fire apparatus approaching. In this situation, the best action for you to take is to
    (A) attempt to clear a path for the fire apparatus by driving rapidly and sounding your horn
    (B) drive to the next intersection and direct traffic until the apparatus has passed
    (C) permit the apparatus to pass and then follow it closely to the fire, sounding your horn as you drive
    (D) pull to the curb, permit the apparatus to pass, and then continue on your way

35. Whenever a public performance given in a theater involves the use of scenery or machinery, a firefighter is assigned to be present. The main reason for this assignment is that
    (A) theaters are located in high-property-value districts
    (B) the use of scenery and machinery increases the fire hazard
    (C) theatrical districts have heavy traffic, making for slow response of apparatus
    (D) emergency exits may be blocked by the scenery or machinery

DIRECTIONS: Questions 36 through 39 are based on the following paragraph.

A plastic does not consist of a single substance; it is a blended combination of several substances. In addition to the resin, it may contain various fillers, plasticizers, lubricants, and coloring material. Depending on the type and quantity of substances added to the binder, the properties, including combustibility, might be altered considerably. The flammability of plastics depends on their composition and, as with other materials, on their physical size and condition. Thin sections, sharp edges, or powdered plastics will ignite and burn more readily than the same amount of identical material in heavy sections with smooth surfaces.

36. According to the paragraph, all plastics contain a
    (A) resin
    (B) resin and a filler
    (C) resin, filler, and plasticizer
    (D) resin, filler, plasticizer, lubricant, and coloring material

37. The conclusion best supported by the paragraph is that the flammability of plastics
    (A) generally is high
    (B) generally is moderate
    (C) generally is low
    (D) varies considerably

38. According to the paragraph, *plastics* can best be described as
    (A) a trade name
    (B) the name of a specific product
    (C) the name of a group of products that have some similar and some dissimilar properties
    (D) the name of any substance that can be shaped or molded during the production process

39. In a manufacturing process, large thick sheets of a particular plastic are cut, buffed, and formed into small tools. The statement most in accord with the information in the preceding paragraph is that
    (A) the dust particles of the plastics are more flammable than the tools or the sheets
    (B) the plastic tools are more flammable than the dust particles or the sheets
    (C) the sheets of the plastic are more flammable than the dust particles or the tools
    (D) there is insufficient information to determine the relative flammability of sheets, tools, and dust particles

DIRECTIONS: *Questions 40 through 42 are based on the following paragraph.*

To guard against overheating of electrical conductors in buildings, an overcurrent protective device is provided for each circuit. This device is designed to open the circuit and to cut off the flow of current whenever the current exceeds a predetermined limit. The fuse, which is a common form of overcurrent protection, consists of a fusible metal element that, when heated by the current to a certain temperature, melts and opens the circuit.

40. According to the paragraph, a circuit that is *not* carrying an electric current is
    (A) an open circuit
    (B) a closed circuit
    (C) a circuit protected by a fuse
    (D) a circuit protected by an overcurrent protective device other than a fuse

41. As used in the paragraph, the best example of a conductor is
    (A) a metal table that comes in contact with a source of electricity
    (B) a storage battery generating electricity
    (C) an electrical wire carrying an electrical current
    (D) a dynamo converting mechanical energy into electrical energy

42. According to the paragraph, the maximum number of circuits that can be handled by a fuse box containing six fuses
    (A) is 3
    (B) is 6
    (C) is 12
    (D) cannot be determined from the information given in the paragraph

DIRECTIONS: *Questions 43 through 45 are based on the following paragraph.*

Unlined linen hose essentially is a fabric tube made of closely woven linen yarn. Due to the natural characteristics of linen, very shortly after water is introduced, the wet threads swell, closing the minute spaces between them and making the tube practically watertight. This type of hose tends to deteriorate rapidly if not thoroughly dried after use or if installed where it will be exposed to dampness or weather conditions. It is not ordinarily built to withstand frequent service or use in which the fabric will be subjected to chafing from rough or sharp surfaces.

43. Seepage of water through an unlined linen hose is observed when the water is first turned on. From the preceding paragraph, we can conclude that the seepage
    (A) indicates that the hose is defective
    (B) does not indicate that the hose is defective provided that the seepage is proportionate to the water pressure
    (C) does not indicate that the hose is defective provided that the seepage is greatly reduced when the hose becomes thoroughly wet
    (D) does not indicate that the hose is defective provided that the seepage takes place only at the surface of the hose

44. Unlined linen hose is most suitable for use
- (A) as a garden hose
- (B) on fire department apparatus
- (C) as emergency fire equipment in buildings
- (D) in fire department training schools

45. The use of unlined linen hose would be *least* appropriate in
- (A) an outdoor lumberyard
- (B) a nonfireproof office building
- (C) a department store
- (D) a cosmetic manufacturing plant

Directions: *Use the information in the following passage to answer questions 46 and 47.*

One of the most common emergency situations that a firefighter might face is a motor vehicle accident. Victims in these situations are commonly trapped in the vehicle, making it difficult for a rescuer to administer proper first aid. It is common, however, for injuries involved in a motor vehicle accident to be very serious and even life threatening. It therefore is important to quickly gain access to the vehicle so that at least one rescuer can be placed in the vehicle to begin stabilizing the victim. After a rescuer gains access to the vehicle, he should conduct an initial survey to assess any life-threatening injuries. While conducting the initial survey, the rescuer should keep the victim's airway open, perform CPR if necessary, and treat any uncontrolled bleeding. These procedures are listed in terms of importance and always should be conducted in that order.

46. It might be difficult for a rescuer to administer first aid to the victim of a motor vehicle accident because the victim often is
- (A) stubborn
- (B) trapped
- (C) breathing
- (D) young
- (E) stabilized

47. The first thing a rescuer should assess after access is gained to the vehicle is
- (A) whether there is uncontrolled bleeding
- (B) whether CPR is necessary
- (C) whether the victim is trapped
- (D) whether any bones are broken
- (E) whether the victim's airway is open

DIRECTIONS: *Use the information in the following passage to answer questions 48 through 50.*

Fire Chief Williams has prepared the following list of procedures, which are not in sequence, for rescuing victims from burning buildings:

1. From the time of entrance, an internal building search always should be conducted on hands and knees. Visibility in a burning building often is poor, making it dangerous to walk upright.

2. Firefighters involved in rescue operations always should enter a burning building in groups of at least two.

3. Victims who have been removed from a building always should be placed in the custody of an individual who can ensure that the victim will not attempt to reenter the building.

4. Full protective clothing and a protective breathing apparatus always should be donned and double-checked for malfunctions prior to any rescue operation.

5. After a room has been thoroughly searched, the firefighter always should leave a sign to other firefighters that the room has been searched. Hang tags should be left on door handles to indicate that the room has been searched.

6. An internal search always should begin with an outside wall. Windows typically are located on the outside wall and can be opened for ventilation.

48. The logical order for the procedures listed above is
    (A) 1, 5, 3, 6, 2, 4
    (B) 6, 5, 4, 2, 3, 1
    (C) 6, 5, 1, 3, 4, 2
    (D) 4, 2, 1, 6, 5, 3
    (E) 4, 6, 2, 1, 5, 3

49. According to the procedures listed, a firefighter who walks inside of a burning building is in violation of which procedural rule?
    (A) 2
    (B) 4
    (C) 5
    (D) 1
    (E) 6

50. The procedures listed imply that tags should be left on the door handles of searched rooms so that
    (A) victims can find their way out of the building
    (B) other firefighters do not search the room again
    (C) victims do not attempt to reenter the building
    (D) windows can be located for ventilation
    (E) protective equipment can be double-checked

51. A firefighter on inspection duty should behave in a manner that is likely to gain the support of the public that his or her agency serves. The firefighter can accomplish this by engaging in all but one of the following practices. The practice that will *not* accomplish this is
   (A) being courteous and neat in appearance
   (B) being prepared to give explanations of the Fire Prevention Code
   (C) overlooking minor infractions of the Fire Prevention Code because correcting them would be expensive
   (D) performing inspection duties in a businesslike manner

52. A firefighter receives a call during the night while off duty from a neighbor who says his house is on fire. The most appropriate action for the firefighter to take *first* is to
   (A) call for the assistance of the fire department if it has not already been called
   (B) go to the neighbor's home to evaluate the extent of the fire
   (C) call other off-duty firefighters who live close by to help fight the fire
   (D) advise the neighbor that you will be over as soon as you are dressed

DIRECTIONS: *Use the information in the following passage to answer questions 53 through 57.*

When several victims are involved in a motor vehicle accident, it is essential that firefighters quickly assess the extent of each victim's injuries so they can be treated and removed from their vehicles in a logical sequence. Victims who are not injured or who are able to exit the vehicle on their own should be quickly removed from the vehicle to make room for firefighters to attend to the most seriously injured individuals. After the most seriously injured individuals have been stabilized, firefighters should give priority to victims trapped in their vehicles.

Last week, firefighters Warren and Mauro were the first to arrive at the scene of a multicar accident that occurred in a rural area. Realizing that the next emergency vehicle would not be able to reach the location for another 20 minutes, the firefighters had to quickly assess which victims should be treated and/or removed from their vehicles.

Three vehicles were involved in the accident. Each vehicle had a driver and one passenger. The first vehicle was severely damaged with the driver trapped inside. The driver of the car was unconscious and did not appear to be breathing. The passenger did not appear to have any life-threatening injuries.

The second car was not damaged as severely. The driver of the car appeared to have a broken leg that was trapped by the steering column but had no other serious injuries. The passenger of the car appeared to be quite shaken from the accident but had no serious injuries.

The third car was the least damaged of the three cars involved in the accident. The driver, however, was not wearing a seat belt and hit the windshield with her head. She had some minor cuts on her forehead and was extremely disoriented. The passenger in her car had a life-threatening compound fracture of his leg that was bleeding profusely.

53. According to the first paragraph, which two victims should be quickly removed from the cars to make room for rescue operations?
    (A) driver #1, passenger #1
    (B) passenger #1, passenger #2
    (C) driver #2, driver #3
    (D) driver #3, passenger #3
    (E) driver #1, passenger #3

54. If firefighter Mauro is treating driver #1 and no other victims have been treated, who should firefighter Warren attend to first?
    (A) passenger #1
    (B) passenger #3
    (C) driver #2
    (D) driver #3
    (E) passenger #2

55. After the most seriously injured individuals have been stabilized, who should firefighters Warren and Mauro attend to?
    (A) driver #1, driver #2
    (B) passenger #1, passenger #2
    (C) driver #1, passenger #1
    (D) driver #2, passenger #2
    (E) driver #3, passenger #1

56. A clear sign that driver #3 might have suffered a major head injury is that
    (A) she had cuts on her forehead
    (B) her car had the least damage
    (C) she did not appear to be breathing
    (D) she was extremely disoriented
    (E) she was trapped in the car

57. Passenger #1 should be quickly removed from the car
    (A) because he/she was seriously injured
    (B) because he/she was bleeding profusely
    (C) because he/she was not injured
    (D) because he/she had a broken leg
    (E) because he/she was trapped in the car

58. The best explanation for why smoke usually rises from a fire is that
    (A) cooler, heavier air displaces lighter, warm air
    (B) heat energy from the fire propels smoke upward
    (C) suction from the upper air pulls the smoke upward
    (D) burning matter is chemically changed into heat energy

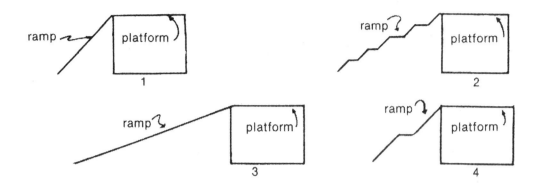

59. The preceding diagram shows various types of ramps leading to a loading platform. The ramp that would permit the load to be moved up to the platform with the least amount of force is
  (A) 1
  (B) 2
  (C) 3
  (D) 4

60. The practice of racing a car engine to warm it up in cold weather generally is
  (A) good, mainly because repeated stalling of the engine and drain on the battery is avoided
  (B) bad, mainly because too much gas is used to get the engine heated
  (C) good, mainly because the engine becomes operational in the shortest period of time
  (D) bad, mainly because proper lubrication is not established rapidly enough

61. Ice on sidewalks often can be melted by sprinkling salt on it. The melting of the ice results from
  (A) a chemical reaction between the salt and the ice that produces heat
  (B) attraction of sun rays by the salt to the ice
  (C) lowering of the freezing point of water by the salt
  (D) the heat of friction caused by persons walking on the salt

62. Only one of the following statements relating to the temperature at which water boils is correct. The correct statement is that
  (A) water always boils at the same temperature regardless of pressure
  (B) water heated slowly by a low flame will boil at a higher temperature than water heated quickly by a high flame
  (C) a large quantity of water will boil at a higher temperature than a small quantity
  (D) water heated at sea level will boil at a higher temperature than water heated on the top of a mountain

63. A substance that is a good conductor of heat is most likely to be a poor
  (A) conductor of electricity
  (B) insulator of heat
  (C) vibrator of sound
  (D) reflector of light

64. At a meeting concerned with fire prevention, this was said: "The fact that fire loss has been maintained near its previous levels is encouraging evidence that our increasing efforts over the years in the field of public education in fire protection have not been unavailing." Of the following, the most essential assumption that must be made if this statement is accepted is that
    (A) further public education in fire protection is desirable
    (B) fire loss has been computed on the basis of real value rather than insured value
    (C) reference is made here to losses due to fires caused by carelessness rather than sabotage
    (D) there has been, in recent years, an increase in the potential fire hazard

65. A load is to be supported from a steel beam by a chain consisting of 20 links and a hook. If each link of the chain weighs 1 pound and can support a weight of 1,000 pounds, and if the hook weighs 5 pounds and can support a weight of 5,000 pounds, the maximum load that can be supported from the hook is most nearly
    (A) 25,000 pounds
    (B) 5,000 pounds
    (C) 1,000 pounds
    (D) 975 pounds

66. Ice formation in water pipes often causes bursting of the pipes because
    (A) the additional weight of ice overloads the pipes
    (B) water cannot pass the ice block and builds up great pressure on the pipes
    (C) the cold causes contraction of the pipes and causes them to pull apart
    (D) water expands upon freezing and builds up great pressure on the pipes

67. The suggestion has been made that groups of firefighters without apparatus of any kind should be kept in reserve at a few centrally located points throughout the city. Of the following, the most valid justification for this proposal is that
    (A) when second or third alarms are sent, the need often is for more firefighters rather than more apparatus
    (B) the fire department is understaffed
    (C) the fire districts in a city should be revised periodically to meet population trends
    (D) discipline is as important as apparatus in extinguishing fires quickly

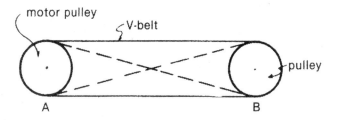

68. In the preceding diagram, crossing the V-belt as shown by the dotted lines will result in
    (A) pulley A reversing direction
    (B) no change in the direction of either pulley
    (C) pulley B reversing direction
    (D) stoppage of the motor

DIRECTIONS: *Questions 69 through 71 are based on the following diagram. Assume that the teeth of the gears are continuous all the way around each gear.*

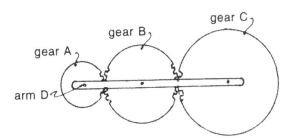

69. Fastening gear A to arm D at another point in addition to its shaft will result in
    (A) gear B rotating on its shaft in a direction opposite to gear A
    (B) gear C rotating on its shaft in a direction opposite to gear A
    (C) arm D rotating around the shaft of gear C
    (D) the locking of all gears

70. If gear C is fastened to a supporting frame (not shown) so that it cannot rotate, and if gear A turns clockwise on its shaft, then
    (A) gear B will turn counterclockwise and arm D will turn clockwise around the shaft of gear C
    (B) gear B will turn counterclockwise and arm D will turn counterclockwise around the shaft of gear C
    (C) gear B will turn clockwise and arm D will turn clockwise around the shaft of gear C
    (D) gear B will turn clockwise and arm D will turn counterclockwise around the shaft of gear C

71. If gear B is fastened to a supporting frame (not shown) so that it cannot rotate, and if arm D rotates clockwise around the shaft of gear B, then for each complete revolution arm D makes, gear A will make
    (A) more than one turn clockwise about its own shaft
    (B) less than one turn clockwise about its own shaft
    (C) more than one turn counterclockwise about its own shaft
    (D) less than one turn counterclockwise about its own shaft

72. As a ship sails away from shore, it appears to go below the horizon. The best explanation for this observation is that it results from the
    (A) rise and fall of tides
    (B) curvature of the earth's surface
    (C) refraction of light
    (D) effect of gravity on moving bodies

73. Of the following, the most important reason for lubricating moving parts of machinery is to
    (A) reduce friction
    (B) prevent rust formation
    (C) increase inertia
    (D) reduce the accumulation of dust and dirt on the parts

74. A canvas tarpaulin measures 6 feet by 9 feet. The largest circular area that can be covered completely by this tarpaulin is a circle with a diameter of
(A) 9 feet
(B) 8 feet
(C) 7 feet
(D) 6 feet

75. A firefighter caught a civilian attempting to reenter a burning building despite several warnings to stay outside of the fire lines. The civilian insisted frantically that he needed to save some very valuable documents from the fire. The firefighter then called a police officer to remove the civilian. The firefighter's action was
(A) wrong; it is bad public relations to order people about
(B) right; the firefighter is charged with the responsibility of protecting lives
(C) wrong; the firefighter should have explained to the civilian why he should not enter the building
(D) right; civilians must be excluded from the fire zone

DIRECTIONS: *Questions 76 through 78 should be answered on the basis of the following diagram. In this diagram, pulley A and pulley B are both firmly attached to the shaft so that both pulleys and the shaft can turn only as a single unit. The radius of pulley A is 4 inches, and the radius of pulley B is 1 inch.*

76. If weight W_A weighs 20 pounds, the system will be in balance if weight W_B weighs
(A) 5 pounds
(B) 10 pounds
(C) 20 pounds
(D) 80 pounds

77. If the rope on pulley A is pulled downward so that it unwinds, the rope on pulley B will
(A) wind slower
(B) wind faster
(C) unwind slower
(D) unwind faster

78. When pulley A makes one complete revolution, pulley B will make
(A) one quarter of a revolution
(B) one revolution
(C) four revolutions
(D) a number of revolutions that cannot be determined from the information given

79. For combustion to take place, the atmosphere must contain at least 12 percent oxygen. The maximum percent of oxygen in an atmosphere that can support combustion is
    (A) 21 percent
    (B) 25 percent
    (C) 50 percent
    (D) 100 percent

80. Firefighters, when not responding to alarms, are assigned to perform housekeeping chores on the equipment in the firehouse. The primary reason for these assignments is to
    (A) keep the firefighters busy when not responding to alarms
    (B) keep the equipment and quarters in the best possible condition
    (C) save money by not hiring maintenance workers
    (D) make firehouses competitive with each other

81. If a firefighter in the act of extinguishing a fire is forced to run through an area of dense smoke, it would be best for the firefighter to
    (A) take deep, slow breaths of air
    (B) breathe deeply before starting and then slowly while passing through the area
    (C) take a shallow breath before starting and a deep breath while passing through the area
    (D) refrain from breathing while passing through the area

82. Empty hoses are dragged to the site of a fire before the water supply is turned on. The reason for this is that
    (A) full hoses are liable to catch on protruding objects
    (B) full hoses are likely to be damaged while being dragged
    (C) the source of the water might be too far removed from the fire
    (D) empty hoses can be dragged more easily and quickly

83. An intense fire develops in a room in which carbon dioxide cylinders are stored. The principal hazard in this situation is that
    (A) the CO_2 might catch fire
    (B) toxic fumes might be released
    (C) the cylinders might explode
    (D) released CO_2 might intensify the fire

84. At a fire involving the roof of a five-story building, the firefighters trained their hose stream on the fire from a vacant lot across the street, aiming the stream at a point about 15 feet above the roof. In this situation, water in the stream would be traveling at the greatest speed
    (A) as it leaves the hose nozzle
    (B) at a point midway between the ground and the roof
    (C) at the maximum height of the stream
    (D) as it drops on the roof

85. Of the following circumstances concerning a fire, the one that indicates most strongly the possibility of arson is that
    (A) there was heavy charting of wood around the point of origin of the fire
    (B) three fires apparently broke out simultaneously in different parts of the building
    (C) the heat was so intense that glass in the building became molten and fused
    (D) when the firefighters arrived the smoke was very heavy

86. A firefighter searching an area after an explosion comes upon a man in great pain with a leg injury bleeding severely. Under these circumstances, the first action the firefighter should take is to attempt to
    (A) immobilize the man's leg
    (B) cover the man with a blanket or coat
    (C) stop the bleeding
    (D) make the man comfortable

87. Spontaneous combustion might be the reason for a pile of oily rags catching fire. In general, spontaneous combustion is the direct result of
    (A) application of flame
    (B) falling sparks
    (C) intense sunlight
    (D) chemical action

88. Most fire departments advocate that firefighters become involved in community activities such as coaching or sponsoring a Little League team. This course of action is likely to
    (A) produce expert Little League coaching
    (B) lead to better understanding of the activities of the fire department
    (C) encourage community members to assist in fighting fires when needed
    (D) enlist community support of the firefighters' quest for higher salaries

89. There is a common bond among all firefighters functioning throughout the United States and even in some other parts of the world. There are many nationwide organizations of firefighters, and it is not uncommon for them to drop in at the local fire department when they are visiting other cities. The most likely reason for this is the
    (A) similarity of techniques, equipment, and problems from one fire department to another
    (B) need for firefighters to be strongly organized to achieve equitable pay
    (C) extreme dangers involved in fighting fires
    (D) usual competitiveness of police and fire departments

90. A firefighter stationed in a firehouse is approached by an irate citizen who declares that firefighters are paid well but spend most of their time lounging around the firehouse doing nothing. It would be most appropriate for the firefighter to respond by saying
    (A) "You had better forward your complaint to the mayor."
    (B) "If you did my job for one week, you would change your opinion."
    (C) "How hard do you work for a living?"
    (D) "The job of a firefighter is to always be ready to respond to an alarm."

91. Firefighters often visit schools to give talks about fire prevention to children. The type of firefighter best suited for this duty is
    (A) one whose demeanor depicts authority but who has little experience in public speaking
    (B) a newly appointed firefighter who is a recent college graduate that majored in speech
    (C) a veteran firefighter who projects a strong image and who has addressed audiences in the past
    (D) the firefighter who has achieved the highest rating on the eligible list

92. During an emergency situation, a firefighter might be the recipient of an order from his or her superior that is delivered in an abrupt manner. The recipient of this type of an order should assume that
    (A) he or she is in disfavor with the superior officer
    (B) the superior officer feels that he or she will not respond properly to any other type of order
    (C) he or she is about to be brought up on charges
    (D) emergency situations call for this type of order

93. Studies of accidents have indicated that more accidents occur at the end of a tour of duty than at its beginning. The most likely reason for this is
    (A) carelessness on the part of the victims
    (B) improper training in accident prevention
    (C) lack of interest in safety
    (D) mental and physical fatigue on the part of firefighters

94. Firefighters can best avoid accidents while doing heavy lifting by
    (A) keeping the knees bent and the back straight
    (B) keeping both knees and the back straight
    (C) using the arms only
    (D) using the method best suited to the individual

95. The *least* likely reason for including an accident-prevention program in the training program for newly appointed firefighters is the
    (A) nature of the work being performed
    (B) number of reported accidents
    (C) number of levels of command
    (D) time lost as a result of accidents

96. A training officer assigned to teach safety to firefighters can best gain their cooperation by
    (A) exercising the authority consistent with his or her rank
    (B) waiting until a serious accident occurs
    (C) making a distinct effort to become "one of the boys" during the training session
    (D) indicating to the trainees how an accident can be costly both to the individual and to the department

97. A veteran firefighter observes a newly appointed firefighter performing a maintenance task in an unsafe manner. It would be most appropriate for the veteran firefighter to
    (A) tell the superior officer of the indiscretion at once
    (B) illustrate the proper manner for performing this task to the rookie firefighter
    (C) not do anything because this is not an assigned responsibility
    (D) tell another rookie firefighter to demonstrate the proper method because it is likely that there is a common bond between them

98. A well-executed accident report is extremely valuable principally because it
    (A) isolates the cause of the accident so that future similar accidents can be avoided
    (B) forms the basis of statistical studies
    (C) complies with existing regulations
    (D) informs everybody that an accident has taken place

99. It is essential that accident reports be submitted as soon after the accident has occurred as is practical. The most important reason for this course of action is that
   (A) witnesses tend to change their stories as time passes
   (B) accident reports should be considered trivial and therefore should be gotten out of the way
   (C) the details of the report will still be fresh in the mind of the reporter
   (D) all reports should be submitted promptly even if they are incomplete

100. Following an accident involving a firefighter on the job, it is normal procedure for a fire department to spend considerable effort in an attempt to identify the cause of the accident. The most important reason for this procedure is
   (A) that no accident report is complete without the cause of the accident
   (B) for statistical purposes
   (C) so that the guilty firefighter can be disciplined
   (D) so that future similar accidents can be avoided

Answer Key for Practice Examination 4

1.	D	21.	C	41.	C	61.	C	81.	D
2.	A	22.	A	42.	D	62.	D	82.	D
3.	D	23.	D	43.	C	63.	B	83.	C
4.	B	24.	A	44.	C	64.	A	84.	A
5.	D	25.	C	45.	A	65.	D	85.	B
6.	C	26.	C	46.	B	66.	D	86.	C
7.	C	27.	C	47.	E	67.	A	87.	D
8.	C	28.	A	48.	D	68.	C	88.	B
9.	B	29.	B	49.	D	69.	D	89.	A
10.	C	30.	D	50.	B	70.	B	90.	D
11.	A	31.	B	51.	C	71.	A	91.	C
12.	C	32.	B	52.	A	72.	B	92.	D
13.	B	33.	B	53.	B	73.	A	93.	D
14.	B	34.	D	54.	B	74.	D	94.	A
15.	C	35.	B	55.	A	75.	B	95.	C
16.	A	36.	A	56.	D	76.	D	96.	D
17.	E	37.	D	57.	C	77.	A	97.	B
18.	D	38.	C	58.	A	78.	B	98.	A
19.	C	39.	A	59.	C	79.	D	99.	C
20.	B	40.	A	60.	D	80.	B	100.	D

Explanatory Answers for Practice Examination 4

1. **(D)** The most effective course of action is to suggest that the manager contact the health department so he is assured of receiving correct information.

2. **(A)** This immediately identifies the person receiving the call and is likely to save time and prevent misunderstandings.

3. **(D)** No area should remain without proper protection to cover possible fires. If a company is busy fighting one fire, coverage should be provided in the event that a second fire occurs in the same area.

4. **(B)** The responding fire company does not know the actual location of the fire. It will respond to the signal received from the box. When the fire apparatus arrives at the location of the box, the firefighter will be able to direct it to the actual location of the fire.

5. **(D)** The badge might fall into the wrong hands and be used improperly. The badge belongs to the fire department. Any found property should be returned to the rightful owner, especially if the owner can be identified.

6. **(C)** Consideration should be given to the fact that as much mail as possible should be salvaged from a mailbox fire. Very often, the contents of the mail are extremely important and valuable.

7. **(C)** One of the difficulties in determining the cause of a fire is that a fire is likely to destroy the evidence of what caused it.

8. **(C)** This device would discourage many people from using fire alarm boxes due to the fear of possible incrimination.

9. **(B)** Water pipes leading to the sprinklers often are blown apart by an explosion.

10. **(C)** This course of action would prevent large amounts of water from entering the subway system.

11. **(A)** Note that the choice that is *not* true is the correct answer. Motorists do not avoid routes of travel because fire apparatus might use them.

12. **(C)** The first action to be taken is to determine whether a fire has been started. The sooner a fire is discovered, the easier it will be to extinguish. This generally is true for all fires.

13. **(B)** Safe driving calls for total concentration. Firefighters should refrain from distracting the chauffeur while he or she is driving the apparatus.

14. **(B)** Most major cities have established a 911 emergency number that accepts all calls and sends assistance for police, fire, and medical emergencies.

15. **(C)** There are 150 hydrants for every square mile. To figure out what percentage of 150 fire hydrants 120 hydrants is, take 120/150 = .80. To convert this to a percent, take 100% × .80 = 80%.

16. **(A)** In one hour, the firefighters would have traveled 30 miles by traveling 30 mph. It will take them a fraction of that time to travel 3.2 miles. Thus, take 3.2/30 = .11. Because there are 60 minutes in an hour, $60 \times .11 = 6.6$ minutes.

17. **(E)** First, do what is in the parentheses, $9 - 7 = 2$. $13 \times 13 = 169$. $169 + 2 = 171$.

18. **(D)** To figure out the equation, simply replace the letters with the corresponding numbers. 500/3.2 − 2(45). First multiply 2 by what is in the parentheses. Then divide 500 by 3.2. Now we have $156.25 - 90 = 66.25$.

19. **(C)** The total number of calls was $21 + 17 + 2 + 46 = 86$. Because station A received 34 calls, this is equal to $34/86 = 40\%$ of the calls.

20. **(B)** The correct spelling is "proficient."

21. **(C)** The correct spelling is "unconscious."

22. **(A)** The word "hidden" is the best word to complete the sentence.

23. **(D)** The word "their" is the best word to complete the sentence. This is the proper possessive form of the word "they."

24. **(A)** The word "raise" is the best word to complete the sentence.

25. **(C)** The fire department accepts liability when it permits a nonmember to ride on its equipment. The decisions to accept this liability should come from headquarters.

26. **(C)** If the pier is damaged during the fighting of the fire, the apparatus might fall from the pier into the water.

27. **(C)** These closed cabs provide greater protection for firefighters from inclement weather and hazardous conditions.

28. **(A)** Fires that occur during severe weather conditions are more difficult to extinguish. Therefore, increased manpower might be needed to combat a fire in a blizzard.

29. **(B)** Bends weaken the hose and therefore should be kept to a minimum.

30. **(D)** The decision as to what to do with the money should come from a higher authority in the fire department.

31. **(B)** A section of a falling cornice might injure a pedestrian or a member of the fire department working on the street level.

32. **(B)** There always is a possibility that a fire might rekindle from the debris and ashes.

33. **(B)** Wetting the firefighters with small streams of water will accomplish nothing but making the firefighters wet.

34. **(D)** There is no reason for you to do anything but get out of the way of the oncoming fire apparatus. Driving rules require all vehicles to pull to the curb to allow fire apparatus to pass.

35. **(B)** Because of the use of scenery, which might be composed largely of wood and other highly combustible materials, and machinery, which might involve electrical wiring or other devices creating a potential fire hazard, a firefighter is assigned to be present to ensure that these materials are handled correctly and that, in the event of a fire, the problem can be remedied as quickly as possible.

36. **(A)** The second sentence indicates this fact.

37. **(D)** This fact is indicated in the third and fourth sentences of the paragraph.

38. **(C)** This is stated in the first, second, and third sentences.

39. **(A)** This can be derived from the last sentence of the paragraph.

40. **(A)** An open circuit does not carry electric current. When a fuse melts due to overheating, an open circuit condition results.

41. **(C)** An electrical wire carrying electrical current is a common example of a conductor.

42. **(D)** There is no information about the ratio of circuits to fuses in this paragraph.

43. **(C)** The second sentence indicates that the introduction of water will cause the threads to swell, thus reducing the space between threads and making the hose practically watertight.

44. **(C)** Unlined linen hose cannot withstand very hard wear; thus, it is best used for emergencies.

45. **(A)** Unlined linen hose could withstand neither the punishment of being dragged over rough lumber nor the exposure to weather conditions.

46. **(B)** The second sentence indicates that victims often are trapped, making it difficult for a rescuer to administer proper first aid.

47. **(E)** The passage indicates that keeping the victim's airway open, performing CPR if necessary, and treating any uncontrolled bleeding are listed in terms of importance. Therefore, a rescuer should first assess whether the victim's airway is open.

48. **(D)** Full protective clothing must be put on before doing anything else. The next step is to enter the building in groups of at least two. After you have entered the building, the search should be conducted on hands and knees. The search then should begin on an outside wall. After a room has been searched, the door should be tagged. Victims that have been removed then should be placed in someone's custody.

49. **(D)** Walking inside a burning building is a violation of procedure #1. An internal building search always should be conducted on hands and knees.

50. **(B)** Procedure #5 indicates that tagging the door of a room is a sign to other firefighters that the room has already been searched. This prevents other firefighters from searching the room again.

51. **(C)** It is not the firefighter's responsibility to decide which laws should be enforced and which should not, regardless of expense. Nonprofessional behavior will not gain respect.

52. **(A)** The sooner the fire department arrives at the scene of a fire, the more likely it is that damage to life and property will be kept to a minimum. A single firefighter would not be able to combat the fire alone without the assistance of other firefighters and the use of firefighting apparatus.

53. **(B)** The first paragraph states that victims who are not injured or who are able to leave the vehicle on their own should be quickly removed from the vehicle to make room for the rescuers. Passenger #1 and Passenger #2 did not have any serious injuries and therefore should be quickly removed from the cars to make room for the rescuers.

54. **(B)** The first paragraph states that the most seriously injured individuals should be stabilized first. If firefighter Mauro is treating driver #1, passenger #3 is the most seriously injured individual left and should be treated next.

55. **(A)** The first paragraph states that, once the most seriously injured individuals have been stabilized, priority should be given to victims trapped in their vehicles. Driver #1 and driver #2 both were trapped in their cars and should be attended to next.

56. **(D)** The cuts on her forehead do not necessarily indicate that she suffered a head injury; therefore, is incorrect. Her extreme disorientation, however, might indicate that she has suffered a head injury.

57. **(C)** The first paragraph states that individuals who are not injured should be quickly removed from their cars.

58. **(A)** The cooler, heavier air, because it is dense, pushes the hot air upward.

59. **(C)** The incline of ramp 3 is not nearly as steep as that of ramp 1. Ramps 2 and 4 have irregularities in their inclines that would make it even more difficult to move a heavy load to the top of the platform.

60. **(D)** Oil pressure should be established before the engine is raced.

61. **(C)** Salt affects ice in this manner.

62. **(D)** Water boils at 212°F at sea level, a lower temperature at lower air pressure. Cooking and baking instructions often vary for certain altitudes above sea level.

63. **(B)** It follows that a substance that carries heat well cannot be used as an insulator of heat.

64. **(A)** If some public education has helped maintain fire losses at even level (costs only go up, so maintaining losses shows the effect of education), then increasing public education should reduce the level of future losses.

65. **(D)** A chain is only as strong as its weakest link. All links in this chain are of equal strength. Each can support 1,000 pounds. The weight of the hook reduces this amount slightly.

66. **(D)** Water expands as it freezes. If the expanding water is in an enclosed area, as in a pipe, the expansion must cause the pipe to burst.

67. **(A)** We first met this principle, the need for extra manpower, in the paragraph for questions 53 through 56.

68. **(C)** When it is desired that one pulley be turned in the opposite direction from its connecting pulley, drive belts must be crossed as shown in the diagram.

69. **(D)** With the center of gear A fastened at one point on the shaft, the gear is able to rotate at that point. If gear A is fastened at an additional point on the shaft, the gear becomes rigid and unable to rotate. It therefore cannot turn gear B, and all gears will lock.

70. **(B)** By anchoring gear C, gear C then becomes the pivoting point. The remaining gears will then rotate around the pivot, which is gear C.

71. **(A)** Gear A, being smaller than gears B and C, must revolve at a greater rate of speed to travel the same distance. Therefore, gear A will make more than one rotation on its own shaft for every revolution that arm D makes.

72. **(B)** Because of the curvature of the earth's surface, a person cannot see beyond the horizon.

73. **(A)** Lubrication with oil, grease, or any other suitable substance will reduce friction between moveable parts of machinery. It is not possible to operate most machinery well without lubrication. In addition, the machinery would wear out quickly because of the friction.

74. **(D)** Because the tarpaulin measures 6 feet by 9 feet, it is not possible to completely cover any circular area that exceeds 6 feet in diameter.

75. **(B)** The firefighter already had given warnings and presumably some explanation to the civilian. The job of the firefighter is to save lives by fighting fires. This firefighter acted correctly in calling a police officer to remove the civilian from danger so that the firefighter could return to fighting the fire.

76. **(D)** There is a 1 to 4 ratio between pulleys B and A. To balance the system, the weight on the smaller pulley must be 4 times the weight on the larger pulley. The weight on the large pulley is 20 pounds; therefore, 4 times 20 pounds equals 80 pounds, which is (D).

77. **(A)** Pulley A has a circumference that is 4 times greater than pulley B. With each revolution, pulley B will be traveling one-fourth the distance traveled by pulley A. This results in one-fourth the amount of rope being wound on the pulley. Therefore, pulley B would wind at a slower rate.

78. **(B)** Both pulleys are attached firmly to the shaft; therefore, both will make the same number of revolutions.

79. **(D)** Oxygen, though noncombustible, will support combustion. An atmosphere of 100 percent oxygen will cause burning material to accelerate with explosive rapidity.

80. **(B)** Housekeeping chores are part of the job of a firefighter. The motivated firefighter will take pride in the condition of the equipment used in the performance of his or her duties.

81. **(D)** By not breathing, the firefighter will expose his or her lungs to a minimum amount of irritation from the smoke.

82. **(D)** Full hoses weigh considerably more than empty hoses because of the weight of the water. Therefore, it is much easier to carry or drag empty hoses to the site of a fire.

83. **(C)** The immediate danger involving the cylinders of CO_2 is that the contents of the cylinders might expand and explode from overheating. Not only will the explosion cause immediate danger to the premises, personnel might be injured from the fragments of metal flying from the exploding cylinders.

84. **(A)** The water pressure will diminish gradually as the water gets farther from the nozzle of the hose.

85. **(B)** It would be most unlikely for three fires to break out simultaneously in the same building from accidental causes. Deliberate setting of the fires must be suspected.

86. **(C)** Whenever severe bleeding occurs, it should be given attention immediately because such bleeding can quickly result in the loss of life.

87. **(D)** Spontaneous combustion is combustion (burning) that occurs from no external cause. Flame and falling sparks are obvious external causes. The heat of intense sunlight might act as a catalyst, speeding up chemical activity in a pile of rags or another hazardous pile of debris, but the actual cause of spontaneous combustion is a chemical reaction within the heap.

88. **(B)** When firefighters become involved in community activities, it is likely that the residents will become more aware of the purpose and objectives of the fire department.

89. **(A)** Firefighters have the same responsibilities throughout the world—to extinguish fires, to save lives, and to minimize property damage. The equipment and techniques used are basically the same throughout the world, although larger cities might have more sophisticated firefighting systems.

90. **(D)** There always will be citizens under the impression that firefighters do not work hard because they often can be seen at the firehouse instead of out fighting fires. It is very important that these citizens be made to understand that, after housekeeping chores are done, the firefighter's prime responsibility is to make himself or herself ready to answer an alarm.

91. **(C)** Firefighting is a serious business; therefore, a proper image of the fire department must be created in the minds of children. The firefighter selected must be skilled in public speaking and must be able to put forth a strong and serious image.

92. **(D)** Emergency situations call for immediate action. Very often, courtesies are brushed over in the transmission of orders. The firefighter should realize that the abrupt manner in which he or she is addressed does not reflect on him or her personally.

93. **(D)** The tired worker is likely to become involved in accidents due to lack of alertness to dangerous situations.

94. **(A)** The weight of the object is evenly distributed with the knees bent and the back straight. No part of the body receives the bulk of the weight.

95. **(C)** The number of levels of command has no bearing on the necessity for such a program.

96. **(D)** The best approach to convince firefighters to take a course in safety and accident prevention seriously is to convince them of how costly an accident can be to themselves as well as to the department.

97. **(B)** The newly appointed firefighter should be corrected at once before an accident occurs and before this method of working becomes a habit.

98. **(A)** Isolating the cause of an accident will alert firefighters to be aware of such situations in the future. This should reduce the chances of a similar accident occurring.

99. **(C)** Time plays tricks on an individual's memory. Therefore, the sooner the report is submitted, the more accurate is it likely to be.

100. **(D)** After an accident occurs, little can be done but suffer its consequences. Similar accidents can be avoided in the future, however, if the cause of such an accident is identified.

Answer Sheet for Practice Examination 5

1. Ⓐ Ⓑ Ⓒ Ⓓ Ⓔ
2. Ⓐ Ⓑ Ⓒ Ⓓ Ⓔ
3. Ⓐ Ⓑ Ⓒ Ⓓ Ⓔ
4. Ⓐ Ⓑ Ⓒ Ⓓ Ⓔ
5. Ⓐ Ⓑ Ⓒ Ⓓ Ⓔ
6. Ⓐ Ⓑ Ⓒ Ⓓ Ⓔ
7. Ⓐ Ⓑ Ⓒ Ⓓ Ⓔ
8. Ⓐ Ⓑ Ⓒ Ⓓ Ⓔ
9. Ⓐ Ⓑ Ⓒ Ⓓ Ⓔ
10. Ⓐ Ⓑ Ⓒ Ⓓ Ⓔ
11. Ⓐ Ⓑ Ⓒ Ⓓ Ⓔ
12. Ⓐ Ⓑ Ⓒ Ⓓ Ⓔ
13. Ⓐ Ⓑ Ⓒ Ⓓ Ⓔ
14. Ⓐ Ⓑ Ⓒ Ⓓ Ⓔ
15. Ⓐ Ⓑ Ⓒ Ⓓ Ⓔ
16. Ⓐ Ⓑ Ⓒ Ⓓ Ⓔ
17. Ⓐ Ⓑ Ⓒ Ⓓ Ⓔ
18. Ⓐ Ⓑ Ⓒ Ⓓ Ⓔ
19. Ⓐ Ⓑ Ⓒ Ⓓ Ⓔ
20. Ⓐ Ⓑ Ⓒ Ⓓ Ⓔ
21. Ⓐ Ⓑ Ⓒ Ⓓ Ⓔ
22. Ⓐ Ⓑ Ⓒ Ⓓ Ⓔ
23. Ⓐ Ⓑ Ⓒ Ⓓ Ⓔ
24. Ⓐ Ⓑ Ⓒ Ⓓ Ⓔ
25. Ⓐ Ⓑ Ⓒ Ⓓ Ⓔ
26. Ⓐ Ⓑ Ⓒ Ⓓ Ⓔ
27. Ⓐ Ⓑ Ⓒ Ⓓ Ⓔ
28. Ⓐ Ⓑ Ⓒ Ⓓ Ⓔ
29. Ⓐ Ⓑ Ⓒ Ⓓ Ⓔ
30. Ⓐ Ⓑ Ⓒ Ⓓ Ⓔ
31. Ⓐ Ⓑ Ⓒ Ⓓ Ⓔ
32. Ⓐ Ⓑ Ⓒ Ⓓ Ⓔ
33. Ⓐ Ⓑ Ⓒ Ⓓ Ⓔ
34. Ⓐ Ⓑ Ⓒ Ⓓ Ⓔ
35. Ⓐ Ⓑ Ⓒ Ⓓ Ⓔ
36. Ⓐ Ⓑ Ⓒ Ⓓ Ⓔ
37. Ⓐ Ⓑ Ⓒ Ⓓ Ⓔ
38. Ⓐ Ⓑ Ⓒ Ⓓ Ⓔ
39. Ⓐ Ⓑ Ⓒ Ⓓ Ⓔ
40. Ⓐ Ⓑ Ⓒ Ⓓ Ⓔ
41. Ⓐ Ⓑ Ⓒ Ⓓ Ⓔ
42. Ⓐ Ⓑ Ⓒ Ⓓ Ⓔ
43. Ⓐ Ⓑ Ⓒ Ⓓ Ⓔ
44. Ⓐ Ⓑ Ⓒ Ⓓ Ⓔ
45. Ⓐ Ⓑ Ⓒ Ⓓ Ⓔ
46. Ⓐ Ⓑ Ⓒ Ⓓ Ⓔ
47. Ⓐ Ⓑ Ⓒ Ⓓ Ⓔ
48. Ⓐ Ⓑ Ⓒ Ⓓ Ⓔ
49. Ⓐ Ⓑ Ⓒ Ⓓ Ⓔ
50. Ⓐ Ⓑ Ⓒ Ⓓ Ⓔ
51. Ⓐ Ⓑ Ⓒ Ⓓ Ⓔ
52. Ⓐ Ⓑ Ⓒ Ⓓ Ⓔ
53. Ⓐ Ⓑ Ⓒ Ⓓ Ⓔ
54. Ⓐ Ⓑ Ⓒ Ⓓ Ⓔ
55. Ⓐ Ⓑ Ⓒ Ⓓ Ⓔ
56. Ⓐ Ⓑ Ⓒ Ⓓ Ⓔ
57. Ⓐ Ⓑ Ⓒ Ⓓ Ⓔ
58. Ⓐ Ⓑ Ⓒ Ⓓ Ⓔ
59. Ⓐ Ⓑ Ⓒ Ⓓ Ⓔ
60. Ⓐ Ⓑ Ⓒ Ⓓ Ⓔ
61. Ⓐ Ⓑ Ⓒ Ⓓ Ⓔ
62. Ⓐ Ⓑ Ⓒ Ⓓ Ⓔ
63. Ⓐ Ⓑ Ⓒ Ⓓ Ⓔ
64. Ⓐ Ⓑ Ⓒ Ⓓ Ⓔ
65. Ⓐ Ⓑ Ⓒ Ⓓ Ⓔ
66. Ⓐ Ⓑ Ⓒ Ⓓ Ⓔ
67. Ⓐ Ⓑ Ⓒ Ⓓ Ⓔ
68. Ⓐ Ⓑ Ⓒ Ⓓ Ⓔ
69. Ⓐ Ⓑ Ⓒ Ⓓ Ⓔ
70. Ⓐ Ⓑ Ⓒ Ⓓ Ⓔ
71. Ⓐ Ⓑ Ⓒ Ⓓ Ⓔ
72. Ⓐ Ⓑ Ⓒ Ⓓ Ⓔ
73. Ⓐ Ⓑ Ⓒ Ⓓ Ⓔ
74. Ⓐ Ⓑ Ⓒ Ⓓ Ⓔ
75. Ⓐ Ⓑ Ⓒ Ⓓ Ⓔ
76. Ⓐ Ⓑ Ⓒ Ⓓ Ⓔ
77. Ⓐ Ⓑ Ⓒ Ⓓ Ⓔ
78. Ⓐ Ⓑ Ⓒ Ⓓ Ⓔ
79. Ⓐ Ⓑ Ⓒ Ⓓ Ⓔ
80. Ⓐ Ⓑ Ⓒ Ⓓ Ⓔ
81. Ⓐ Ⓑ Ⓒ Ⓓ Ⓔ
82. Ⓐ Ⓑ Ⓒ Ⓓ Ⓔ
83. Ⓐ Ⓑ Ⓒ Ⓓ Ⓔ
84. Ⓐ Ⓑ Ⓒ Ⓓ Ⓔ
85. Ⓐ Ⓑ Ⓒ Ⓓ Ⓔ
86. Ⓐ Ⓑ Ⓒ Ⓓ Ⓔ
87. Ⓐ Ⓑ Ⓒ Ⓓ Ⓔ
88. Ⓐ Ⓑ Ⓒ Ⓓ Ⓔ
89. Ⓐ Ⓑ Ⓒ Ⓓ Ⓔ
90. Ⓐ Ⓑ Ⓒ Ⓓ Ⓔ
91. Ⓐ Ⓑ Ⓒ Ⓓ Ⓔ
92. Ⓐ Ⓑ Ⓒ Ⓓ Ⓔ
93. Ⓐ Ⓑ Ⓒ Ⓓ Ⓔ
94. Ⓐ Ⓑ Ⓒ Ⓓ Ⓔ
95. Ⓐ Ⓑ Ⓒ Ⓓ Ⓔ
96. Ⓐ Ⓑ Ⓒ Ⓓ Ⓔ
97. Ⓐ Ⓑ Ⓒ Ⓓ Ⓔ
98. Ⓐ Ⓑ Ⓒ Ⓓ Ⓔ
99. Ⓐ Ⓑ Ⓒ Ⓓ Ⓔ
100. Ⓐ Ⓑ Ⓒ Ⓓ Ⓔ

TEAR HERE

PRACTICE EXAMINATION 5

The time allowed for the entire examination is 3 hours and 55 minutes.

DIRECTIONS: Each question has four suggested answers lettered A, B, C, and D. Decide which one is the best answer. On the sample answer sheet, find the question number that corresponds to the question you have just read and darken the circle of the letter you have selected. (Note: This is an examination measuring how well you read, think, and reason. It is not a test of your firefighting knowledge nor is it a course of instruction. Some questions spell out an order of action. You must accept the information as given in the question, even if you doubt its accuracy.)

Questions 1 through 18 should be answered on the basis of the following sketch. You will have five minutes to study the sketch. You then must answer the 18 questions without referring back to it.

Turn the page and study the sketch for exactly five minutes. After five minutes, turn the page and proceed to the 100 questions. You will have 3 hours and 50 minutes to answer the questions. Do not turn back to the sketch after you have begun the questions.

1. The vehicles involved in the motor vehicle accident are
    (A) a taxi and a bus
    (B) a taxi and a truck
    (C) a passenger car and a truck
    (D) a passenger car and a taxi

2. From studying the sketch, one could assume that the accident was caused by the
    (A) taxi
    (B) truck
    (C) bus
    (D) police car

3. The police officer not wearing a hat is
    (A) standing next to the vehicles involved in the accident
    (B) directing traffic at the intersection
    (C) helping a person into the ambulance
    (D) talking to a group of people on the corner

4. The police officer wearing a hat is
    (A) standing next to the tow truck
    (B) standing next to the vehicles involved in the accident
    (C) helping a person into the ambulance
    (D) directing traffic at the intersection

5. The fire is
    (A) at Tony's Pizza
    (B) at 26 Washington Avenue
    (C) above John's Drug Store
    (D) above Tony's Pizza

6. Of the following vehicles, the one *not* depicted in the sketch is
    (A) a motorcycle
    (B) a taxi
    (C) a garbage truck
    (D) a bus

7. The accident occurred in front of
    (A) Tony's Pizza
    (B) John's Drug Store
    (C) Public School #11
    (D) an apartment house

8. The quickest *legal* route by which a fire engine could travel to the scene of the fire is
    (A) west on Washington Avenue
    (B) north on Lincoln Street
    (C) east on Washington Avenue
    (D) south on Lincoln Street

9. The police car is parked
   (A) on Washington Avenue
   (B) on Lincoln Street
   (C) on Washington Street
   (D) in the middle of the intersection of Washington Avenue and Lincoln Street

10. The quickest, most convenient access to the fire scene is to approach from the
    (A) west on Washington Avenue
    (B) south on Washington Street
    (C) north on Lincoln Street
    (D) north on Washington Street

11. The person carrying a large portable radio is standing next to
    (A) Tony's Pizza
    (B) John's Drug Store
    (C) the police car
    (D) the tow truck

12. The ambulance shown in the sketch is from
    (A) Jefferson Hospital
    (B) Columbia Hospital
    (C) Lincoln Hospital
    (D) Washington Hospital

13. The type of establishment at 30 Washington Avenue is
    (A) a gift shop
    (B) a boxing studio
    (C) a drug store
    (D) not indicated

14. A woman with a baby carriage is
    (A) in front of John's Drug Store
    (B) on the corner of Washington Street and Lincoln Street
    (C) in front of Tony's Pizza
    (D) in front of Public School #11

15. A fire alarm box is
    (A) located on the corner of Washington Avenue and Lincoln Street
    (B) located in front of Tony's Pizza
    (C) located in front of Public School #11
    (D) not shown in the sketch

16. The person with the best view of the fire
    (A) is looking out a window
    (B) has long hair
    (C) is near the ambulance
    (D) is driving a tow truck

17. Parking is prohibited on
    (A) Washington Avenue between 8 AM and 7 PM
    (B) Lincoln Street between 7 AM and 8 PM
    (C) Washington Avenue between 8 AM and 7 PM on Sundays
    (D) Washington Street between 7 AM and 8 PM weekdays

18. The fire hydrant closest to the fire is
    (A) in front of Public School #11
    (B) at the southeast corner of Washington Avenue
    (C) at the northwest corner of Washington Avenue
    (D) next to a trash basket

DIRECTIONS: *Answer questions 19 and 20 by selecting the paragraph in which the sentences appear in the most logical order and make the best sense.*

19. (A) You may admire Joe Montana, Louis Nizer, or a good homicide detective. But this does not mean you can count on being successful or happy as a professional ball player, criminal lawyer, or famous detective. It is risky to choose an occupation just because you admire or are fond of someone who has chosen it. Sometimes young people make the mistake of picking a job just because a much-admired relative or friend likes that job.
    (B) Sometimes young people make the mistake of picking a job just because a much-admired relative or friend likes that job. But this does not mean you can count on being successful or happy as a professional ball player, criminal lawyer, or famous detective. You may admire Joe Montana, Louis Nizer, or a good homicide detective. It is risky to choose an occupation just because you admire or are fond of someone who has chosen it.
    (C) Sometimes young people make the mistake of picking a job just because a much-admired relative or friend likes that job. It is risky to choose an occupation just because you admire or are fond of someone who has chosen it. But this does not mean you can count on being successful or happy as a professional ball player, criminal lawyer, or famous detective. You may admire Joe Montana, Louis Nizer, or a good homicide detective.
    (D) Sometimes young people make the mistake of picking a job just because a much-admired relative or friend likes that job. It is risky to choose an occupation just because you admire or are fond of someone who has chosen it. You may admire Joe Montana, Louis Nizer, or a good homicide detective. But this does not mean you can count on being successful or happy as a professional ball player, criminal lawyer, or famous detective.

20. (A) Together, they can fashion a program that will foster public understanding and enlist public assistance in combating crime. This important service makes private security a natural ally of the police and a formidable foe of the criminal. Both police and private security stand to gain, but more importantly, the public stands to gain the most. The business of the private-security sector is not only to sell safety and security but to educate people in the many ways they can protect themselves.
    (B) The business of the private-security sector is not only to sell safety and security but to educate people in the many ways they can protect themselves. Together, they can fashion a program that will foster public understanding and enlist public assistance in combating crime. This important service makes private security a natural ally of the police and a formidable foe of the criminal. Both police and private security stand to gain, but more importantly, the public stands to gain the most.

(C) The business of the private-security sector is not only to sell safety and security but to educate people in the many ways they can protect themselves. Both police and private security stand to gain, but more importantly, the public stands to gain the most. Together, they can fashion a program that will foster public understanding and enlist public assistance in combating crime. This important service makes private security a natural ally of the police and a formidable foe of the criminal.

(D) The business of the private-security sector is not only to sell safety and security but to educate people in the many ways they can protect themselves. This important service makes private security a natural ally of the police and a formidable foe of the criminal. Together, they can fashion a program that will foster public understanding and enlist public assistance in combating crime. Both police and private security stand to gain, but more importantly, the public stands to gain the most.

21. A firefighter might be required to administer first aid to people who are injured. It is of utmost importance that the firefighter be cognizant of certain conditions while performing this function. Of the following, which is the one condition he or she should *not* be concerned with?
    (A) Are the injured person's legs twisted in a manner that might indicate that bones are broken?
    (B) Is the injured party breathing?
    (C) Is there an open medication bottle nearby?
    (D) Is it an apparent case of attempted suicide?

22. Firefighters at the scene of a fire have just about succeeded in putting out the fire. Before it has been completely extinguished, the couple who lives on the top floor requests permission to enter their apartment to secure some of their belongings. It would be most proper for the firefighter who receives this request to
    (A) permit the couple to enter their apartment because the fire is almost out
    (B) suggest that the couple give him a written list and he will get whatever the couple wanted to remove
    (C) permit one but not both persons to return to the apartment
    (D) inform them that they will be permitted to enter only when the building is considered completely safe

DIRECTIONS: *Answer questions 23 and 24 by choosing which implement in the lettered boxes usually is used with the implement or object on the left. Note: The tools are not drawn to scale.*

23.

24.

25. When responding to a fire in a nursing home, extra companies should be dispatched, even on a first alarm. The officer in charge of the first ladder company to arrive must size up the fire quickly to determine whether there will be any need to evacuate residents. If evacuation will be extensive, the officer must immediately call for further reinforcements and must ask the nursing home administrator to direct firefighters to those areas in which residents will need assistance. Simultaneously with the evacuation of residents, steps must be taken to extinguish the fire.

    A frantic nurse has reported a fire in the laundry room of the west wing of the Happy Hours Nursing Home at West 118th Street. The fire dispatcher has sent four pieces of apparatus to the scene. Captain Villegas has sized up the fire and has determined that the fire can be contained and extinguished without any threat to the residents. Now Captain Villegas should
    (A) ask the administrator where the most disabled residents are located
    (B) supervise the fire extinguishing operation
    (C) call for reinforcements
    (D) assist the firefighters in evacuating residents

26. Safety regulations require certain precautions to be taken when pulling fire vehicles out of the firehouse, regardless of the urgency of the fire. Sirens and flashing lights on both sides of the engine house doors must be activated before any vehicle is moved; doors may open at the same time. Then a firefighter should run outside to make sure that no illegally parked cars or trucks are blocking the entrance or turning area and that the street is not congested. Another firefighter should control pedestrian traffic, and one more should stand in the street to detain oncoming traffic until the firefighting equipment has departed.

    The alarm has sounded at Ladder Company 41, and the doors of the engine house are open. The siren is sounding and the lights are flashing. Firefighter Feeney has shouted and signaled that the coast is clear. Firefighter Moro is standing in the street to hold back oncoming cars. Firefighter Williams should
    (A) drive out the first fire engine
    (B) move forward with the first piece of equipment
    (C) step outside and detain pedestrians until all danger has passed
    (D) close the doors to the firehouse

27. If you are asked how to open a fire hydrant, the best way to give this information is to say
    (A) turn the big nut to the left
    (B) turn the big side nut counterclockwise
    (C) turn the big side nut to the left
    (D) turn the big side nut to your left

28. There is a gas leak in the street, and firefighters must be certain that gas is cut off from each building before repairs can be completed. You will help the firefighters best by saying that the gas shut-off in your building is
    (A) in the basement next to the electric meter
    (B) in the far corner of the lower level
    (C) downstairs and to your right
    (D) in the west corner of the basement

DIRECTIONS: *Questions 29 through 31 are based on the map below. The flow of traffic is indicated by the arrows. If only one arrow is shown, traffic flows only in the direction indicated by the arrow. If two arrows are shown, traffic flows in both directions. You must follow the flow of traffic.*

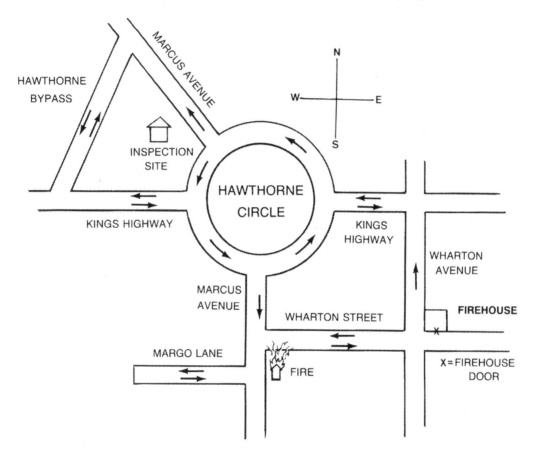

29. A single fire truck leaves the firehouse to make a scheduled fire safety inspection of a building on the northern spur of Marcus Avenue. The best way to get there is to travel
    (A) north on Wharton Avenue, turn left onto Kings Highway, go clockwise around Hawthorne Circle, and exit the circle onto Marcus Avenue just north of Kings Highway
    (B) west on Wharton Street, turn right onto Marcus Avenue, enter Hawthorne Circle going counterclockwise, pass one Kings Highway exit, and exit onto Marcus Avenue
    (C) west on Wharton Street and immediately north on Wharton Avenue, turn left to go west on Kings Highway, turn right onto Hawthorne Circle to the next intersecting road, and turn right onto Marcus Avenue
    (D) north on Wharton Street, west on Kings Highway into Hawthorne Circle, and then counter-clockwise on Hawthorne Circle directly onto Marcus Avenue

30. While the inspection is in progress, the alarm sounds for a fire on the southern spur of Marcus Avenue. The firefighters interrupt their inspection, get on the truck, and proceed to the fire by traveling
    (A) southeast on Marcus Avenue, turning right onto Hawthorne Circle, continuing past Kings Highway, and turning right onto Marcus Avenue
    (B) northwest on Marcus Avenue, making a sharp left turn onto the Hawthorne Bypass, making another left onto Kings Highway, making a right onto Hawthorne Circle, and making the next right to Marcus Avenue

(C) northwest on Marcus Avenue, making a sharp left turn onto the Hawthorne Bypass, making a right onto Kings Highway, making a right onto Hawthorne Circle, and making the next right to Marcus Avenue

(D) southeast on Marcus Avenue to Hawthorne Circle, going clockwise around Hawthorne Circle, passing Kings Highway to the south exit for Marcus Avenue

31. Two engines respond to this same fire directly from the fire station. They travel
    (A) south on Wharton Street and turn left onto Marcus Avenue
    (B) north on Wharton Avenue, west on Kings Highway, and counterclockwise on Hawthorne Circle past the northern spur of Marcus Avenue and the western extension of Kings Highway to the southern spur of Marcus Avenue
    (C) west on Wharton Street and south on Marcus Avenue
    (D) north on Wharton Avenue, west on Kings Highway, and clockwise around Hawthorne Circle to Marcus Avenue

DIRECTIONS: *Answer questions 32 and 33 by selecting the paragraph in which the sentences appear in the most logical order and make the best sense.*

32. (A) They observed a vehicle accident with several injured persons lying on the freeway. The firefighters immediately requested that an ambulance be dispatched to the scene. Prompt action was taken to establish traffic control to protect the injured parties from further injury. Two firefighters returning to the station in the chief's car were traveling on a local freeway.
    (B) Two firefighters returning to the station in the chief's car were traveling on a local freeway. They observed a vehicle accident with several injured persons lying on the freeway. The firefighters immediately requested that an ambulance be dispatched to the scene. Prompt action was taken to establish traffic control to protect the injured parties from further injury.
    (C) Prompt action was taken to establish traffic control to protect the injured parties from further injury. The firefighters immediately requested that an ambulance be dispatched to the scene. They observed a vehicle accident with several injured persons lying on the freeway. Two firefighters returning to the station in the chiefs car were traveling on a local freeway.
    (D) Two firefighters returning to the station in the chief's car were traveling on a local freeway. The firefighters immediately requested that an ambulance be dispatched to the scene. They observed a vehicle accident with several injured persons lying on the freeway. Prompt action was taken to establish traffic control to protect the injured parties from further injury.

33. (A) As contradictory as it might sound, the slower a person reads, the less he absorbs. The faster he reads, the more he understands and retains. This is probably because heavier concentration is required for rapid reading, and concentration enables a reader to grasp important ideas in the reading material. The two basic elements in reading interpretation are speed and comprehension.

(B) The two basic elements in reading interpretation are speed and comprehension. As contradictory as it might sound, the slower a person reads, the less he absorbs. This is probably because heavier concentration is required for rapid reading, and concentration enables a reader to grasp important ideas in the reading material. The faster he reads, the more he understands and retains.

(C) As contradictory as it might sound, the slower a person reads, the less he absorbs. This is probably because heavier concentration is required for rapid reading, and concentration enables a reader to grasp important ideas in the reading material. The faster he reads, the more he understands and retains. The two basic elements in reading interpretation are speed and comprehension.

(D) The two basic elements in reading interpretation are speed and comprehension. As contradictory as it might sound, the slower a person reads, the less he absorbs. The faster he reads, the more he understands and retains. This is probably because heavier concentration is required for rapid reading, and concentration enables a reader to grasp important ideas in the reading material.

DIRECTIONS: *Questions 34 through 36 are based on the following storefront. In this drawing, the storefront is viewed from across the street.*

34. Which of the following drawings shows the same store window at the same time but as seen from inside the store?

35. Which of the following drawings shows the same store window at the same time but as seen from an aerial ladder 17 feet above the street?

36. Which of the following drawings shows the same store window at the same time but as seen through the open trap door in the ceiling at the front of the store?

37. At a house fire, a woman is standing on the lawn screaming, "My baby, my baby!" The best instruction to give to the firefighters is to tell them to
    (A) look for a baby
    (B) search very quietly so they can listen for crying
    (C) systematically check every room for a crib
    (D) start by searching rooms with "tot finder" stickers on the windows

38. The best way to explain why you strongly suspect that the cause of a fire was arson is to say that
    (A) the floor was wet
    (B) the place smelled of gasoline
    (C) there was a funny smell
    (D) you smelled gas

DIRECTIONS: *Use the information in the following passage to answer questions 39 and 40.*

Engine 36 was dispatched to 44 Larson Rd. at 11:32 PM. The engine arrived on the scene at 11:35 PM. Firefighters Wilson and Fitch were greeted at the door of the residence by a distraught man in his mid-thirties. The man explained to the firefighters that his wife was in labor and that he was afraid that taking her to the hospital on his own would be unsafe.

The man led the firefighters into the bedroom where his wife was laying on the bed. She appeared to be in good condition but was very anxious. Wilson asked the woman if she was in severe pain and checked her vital signs. The woman said she was in pain but that it was bearable and that her contractions were about 12 minutes apart. Wilson instructed the woman to do her best to relax and told her that an ambulance would arrive shortly to take her to the hospital.

At 11:39 PM, the ambulance arrived on the scene. The EMT personnel brought a stretcher into the bedroom. The two of them, along with the two firefighters, lifted the woman on to the stretcher. As she was brought out to the ambulance, Fitch assured the woman's husband that everything seemed to be going well and that he had made the right choice by calling 911. Fitch then helped the man into the ambulance so he could ride to the hospital with his wife. The firefighters left the scene at 11:47 PM.

39. The person who checked the woman's vital signs was
    (A) Firefighter Fitch
    (B) the EMT personnel
    (C) the woman's husband
    (D) Firefighter Wilson
    (E) the doctor

40. The length of the call was
    (A) 5 minutes
    (B) 7 minutes
    (C) 10 minutes
    (D) 13 minutes
    (E) 15 minutes

41. An off-duty firefighter sprained an ankle while playing football. The next day, the firefighter returned to work even though the ankle was quite painful. The firefighter felt he was needed on the job. This course of action is considered improper mainly because the firefighter should have known that
    (A) no one is indispensable
    (B) medical attention was more necessary than reporting for duty
    (C) because of this injury, it might be impossible to perform his share of the work when required at a fire
    (D) because the injury occurred off duty, the medical expenses should not be borne by the fire department

42. While fighting a fire, a firefighter searches an apartment looking for victims and comes across an unconscious woman on a bed. The firefighter would probably be of most assistance to the woman by first
    (A) calling for a doctor
    (B) giving the woman first aid until she regains consciousness
    (C) moving the woman to a safe place
    (D) remaining with the woman in case she regains consciousness

DIRECTIONS: *Answer questions 43 and 44 by choosing which implement in the lettered boxes usually is used with the implement on the left. Note: The tools are not drawn to scale.*

43.

44.

DIRECTIONS: *Use the information in the following passage to answer questions 45 through 48.*

In early December, Ed decided to inspect the lake behind his home to see whether the ice was thick enough for him to start ice fishing. The ice close to the shore was very thick, so Ed felt the lake was safe to walk on. He then decided to walk towards the center of the lake. Ed did not realize that the ice on the lake was very thin towards the center, and as he walked out onto the lake, he fell through the ice about 30 feet from the shore. Fortunately, a neighbor saw Ed fall through the ice and immediately called 911.

Minutes later, a fire engine arrived on the scene. Two firefighters brought a wooden ladder with them to the shore near where Ed had fallen into the water. They could see that he was above the water but was not able to pull himself onto the ice. The firefighters asked Ed his name and if he could understand them. Ed did not respond, so the firefighters knew he was in the early stages of hypothermia and needed to be removed from the water as fast as possible. His body temperature was dropping rapidly, and he soon would be unable to assist in his own rescue.

The firefighters took the wooden ladder and slid it on top of the ice towards where Ed had fallen in. When it was close enough, they instructed Ed to grab onto the ladder. He did so and the firefighters pulled Ed onto the ice and dragged him to safety. By the time Ed was on the shore, an ambulance had arrived. It took him to the hospital to be treated for hypothermia.

45. Once Ed was on the shore, the rescuers were most concerned with
    (A) whether Ed had broken any bones
    (B) treating Ed's hypothermia
    (C) whether Ed was able to speak
    (D) treating Ed's frostbite
    (E) pulling Ed out of the water

46. The firefighters knew that Ed was in the early stages of hypothermia because
    (A) he did not respond to the firefighters
    (B) he fell in 30 feet from the shore
    (C) his neighbor had to call 911
    (D) he was able to grab onto the ladder
    (E) the ambulance driver told them

47. It is likely that the firefighters used a ladder to reach Ed because
    (A) his body temperature was dropping
    (B) the ambulance had not yet arrived
    (C) he was unable to grab the ladder
    (D) the ice near Ed was very thin
    (E) the ice was safe to walk on

48. It was important for Ed to be removed from the water quickly because
    (A) the ambulance was waiting on shore
    (B) Ed was not able to swim
    (C) the firefighters were getting cold
    (D) the ice was starting to melt
    (E) Ed would soon be unable to help himself

49. If you are on housewatch and receive the call "Fire, Fire," the best way to get useful information is to ask
    (A) Where's the fire?
    (B) What is the street address of the fire?
    (C) Where are you?
    (D) What is the name of the building on fire?

50. If you are asked when an alarm was received at the firehouse, your best answer would be
    (A) four minutes before the first ladder company arrived at the scene
    (B) just as the night shift was coming on
    (C) at 8:03 PM
    (D) an hour and a half before the fire was declared under control

DIRECTIONS: *Use the following map to answer questions 51 through 53. The flow of traffic is indicated by arrows. If only one arrow is shown, traffic flows only in the direction indicated by the arrow. If two arrows are shown, traffic flows in both directions. You must follow the flow of traffic.*

51. An alarm has just come into the firehouse from the alarm box at the corner of State Street and Grand Boulevard. The quickest route for the fire engines to take is
    (A) out the east exit of the firehouse, north on High Street, south on Grand Street, west on Highland Boulevard to State Street, and south on State Street to the fire
    (B) out the 18th Avenue exit of the fire station and west to 18th Street, north on 18th Street to Boulevard, east on Boulevard to Grand Street, south on Grand Street to Highland Boulevard, west on Highland Way to State Street, and south on State Street to the fire
    (C) out the west exit of the fire station and up 17th Street to Boulevard, two blocks west on Boulevard to State Street, and all the way south on State Street to Grand Boulevard
    (D) out the south exit of the fire station, west on 18th Avenue to State Street, and then south on State Street to its corner with Grand Boulevard

52. A firefighter has been injured fighting this fire and must be transported by ambulance to City Hospital. The hospital emergency exit is located on 17th Street. The hospital is nearby, but it is not easy to get to from the corner of State Street and Grand Boulevard. The best way to go is
  (A) Grand Boulevard to High Street, left onto High Street, left onto Highland Boulevard, and right onto 17th Street
  (B) Grand Boulevard to High Street, left onto High Street, right onto Highland Way, left onto Grand Street, left onto 18th Avenue, and left again onto 17th Street
  (C) east on Grand Boulevard to Highland Boulevard, west on Highland Boulevard to Highland Way, east on Highland Way, and south on 17th Street to the emergency room
  (D) east on Grand Boulevard to Grand Street, left onto Grand Street, left onto 18th Avenue, and south on 17th Street

53. A fire engine is about to leave the station house to make a routine inspection of a building located on the southwest corner of High Street and Highland Boulevard. The best route for the engine to take is to go
  (A) out the High Street exit of the fire station and take High Street straight to Highland Boulevard
  (B) out the east exit of the fire station and up to Boulevard, take Boulevard to Grand Street, Grand Street to Highland Boulevard, and Highland Boulevard to High Street
  (C) out the west exit of the fire station and straight down 17th Street to Highland Boulevard, taking Highland Boulevard to its corner with High Street
  (D) out the 18th Avenue exit of the firehouse to 17th Street, going south on 17th Street to Highland Way and the corner of High Street

DIRECTIONS: Use the information in the following passage to answer questions 54 through 57.

Flaming combustion can be stopped by removing at least one of the four essential elements of the combustion process. This type of combustion can be stopped by reducing the temperature, removing the fuel, removing the oxygen, and stopping the chain reaction involved in flaming.

The most common way to extinguish a flaming fire is to cool the temperature of the fuel with water. By adequately reducing its temperature, a fuel no longer will produce sufficient vapor to burn.

Some fires can be extinguished by removing the fuel source of the fire. If it is a gas or liquid fuel fire, extinguishing it might simply consist of turning off the fuel source. Allowing the fire to consume all the available fuel or letting the fire "burn out" also might be effective.

A flaming fire also can be extinguished by removing the oxygen available to the fire. This can be accomplished by flooding the area around the fire with a nonflammable gas such as carbon dioxide, which will displace the oxygen. A blanket of extinguishing foam also can separate the fuel from the available oxygen.

Stopping the chemical reaction involved in flaming also can effectively extinguish a fire. Some dry chemical agents such as halons can interrupt the chemical reaction that produces flaming.

54. The four essential elements involved in flaming combustion are
 (A) flaming, combustion, fuel, oxygen
 (B) combustion, fuel, chemical reaction, oxygen
 (C) temperature, fuel, oxygen, chemical reaction
 (D) temperature, fuel, oxygen, combustion
 (E) chemical reaction, fuel, flaming, oxygen

55. If a pile of wooden building supplies were on fire, the most common method for extinguishing the fire would be
 (A) cooling it with water
 (B) removing the fuel
 (C) removing the oxygen
 (D) stopping the chemical reaction
 (E) removing the elements

56. To extinguish a flaming fire, it is necessary to remove at least _____ of the essential elements of the combustion process.
 (A) 5
 (B) 4
 (C) 3
 (D) 2
 (E) 1

57. It is possible to interrupt the chemical reaction involved in flaming with
 (A) halons
 (B) heat
 (C) oxygen
 (D) fuel
 (E) "burn out"

DIRECTIONS: *Questions 58 through 60 are based on the following boat collision scene. In this drawing, the scene is viewed from a helicopter coming to the rescue that is just approaching the rear end of a barge from above the yacht with which it collided.*

58. Which of the following drawings shows the same scene at the same time as the original drawing but from the viewpoint of a bystander standing on the shore?

59. Which of the following drawings shows the same scene at the same time as the original drawing but as seen by firefighters about to launch a fireboat from the opposite side of the river?

60. Which of the following drawings shows the same scene at the same time as the original drawing but as seen from above the tugboat?

61. Firefighters responding to an alarm find a badly beaten and unconscious man lying next to the alarm box. It would be appropriate for the firefighters to do all of the following *except*
    (A) summon the police
    (B) give first aid to the man
    (C) summon an ambulance
    (D) question the homeowners in the vicinity

62. During the times when firefighters are not responding to alarms, they busy themselves checking the equipment and vehicles. The most likely reason for them to engage in this activity is to
    (A) be busy when the public visits their firehouse
    (B) provide free time to socialize
    (C) promote neighborhood pride in the appearance of the firehouse
    (D) ensure that their equipment and vehicles are in good working order

DIRECTIONS: *Answer questions 63 and 64 by choosing which implement in the lettered boxes usually is used with the implement on the left. Note: The tools are not drawn to scale.*

63.

64.

DIRECTIONS: *Use the following table to answer questions 65 through 67.*

The following is a table of the decreases in fires from 1994 to 1995 in cities that encouraged fire prevention among their citizens after 1994.

NUMBER OF FIRES IN 1994/1995

City	1994	1995	Decrease
Arlington	427	374	53
Tolbury	12	9	3
Burlington	532	465	67
Littleton	127	132	-5
Deveron	86	63	23
Total	**1184**	**1043**	**141**

65. The number of fires in Deveron in 1995 was ____.
    (A) 109
    (B) 98
    (C) 72
    (D) 63
    (E) 55

66. The city with the greatest percentage decrease in fires from 1994 to 1995 was _____.
    (A) Deveron
    (B) Tolbury
    (C) Littleton
    (D) Arlington
    (E) Burlington

67. The ratio of fires from 1994 to 1995 for the city of Tolbury was _____.
    (A) 4:3
    (B) 2:1
    (C) 1:2
    (D) 3:4
    (E) 5:4

68. When a hose is connected to a fire hydrant and one valve is opened, the water fills the hose at a rate of 5 feet per second. When two hoses are connected to a fire hydrant and both valves are open, each hose fills with water at a rate of 2 feet per second. If the valve for hose A is turned on 3 seconds before hose B and both hoses are 125 feel long, when will hose A fill completely with water?
    (A) after 25 seconds
    (B) after 62 seconds
    (C) after 58 seconds
    (D) after 30 seconds
    (E) after 65 seconds

69. At temperatures of 212°F, a cubic foot of water can expand to about 1,700 times its original volume. If four gallons of water are heated at 212°F, what is the area that the resulting steam will occupy?
    (A) 5,000 cubic feet
    (B) 5,800 cubic feet
    (C) 6,300 cubic feet
    (D) 6,500 cubic feet
    (E) 6,800 cubic feet

70. At the scene of an apartment house fire, the persons in the greatest danger from the fire must be rescued first. Thus, occupants of the room on fire must be removed first, followed by occupants of the remainder of the fire floor. Heat rises, so occupants of the floors above the fire floor generally are in greater danger than occupants of floors below the fire floor. Occupants of the floor directly above the fire floor have the next priority for evacuation. Although the general rule is that the floors above the fire present greater danger, the one exception is the floor directly below the fire. Because water is very heavy and fire tends to undermine the structure of a building, the fire floor might collapse onto the floor beneath. People on the floor directly below run the risk of being crushed by the floor and furniture from above. The floor below the fire, then, should be the third to be emptied. As long as the fire is actively burning, upper floors have priority for further evacuation. After the fire has been declared under control, evacuation should proceed on lower floors because of the continued danger from water damage.

The fire on the fourth floor of a block-deep six-story apartment house has just been declared under control. Occupants of the fourth floor were all evacuated by way of the fire escape and ladders. Some fifth-floor occupants have come down the fire escape; others have made their way to the roof with some sixth-floor residents. Firefighters should now concentrate on evacuating

(A) the third floor
(B) the second floor
(C) the sixth floor
(D) the roof

DIRECTIONS: *Questions 71 through 73 are based on the following airplane disaster. In this drawing, the scene is viewed from the driver's seat of a fire engine approaching the right side of the airplane near the tail.*

71. Which of the following drawings shows the same scene at the same time but as seen from a rescue launch approaching the nose of the plane?

72. Which of the following drawings shows the same scene at the same time but as seen from a news helicopter hovering directly overhead?

73. Which of the following drawings shows the same scene at the same time but as seen by a person running up from behind the tail?

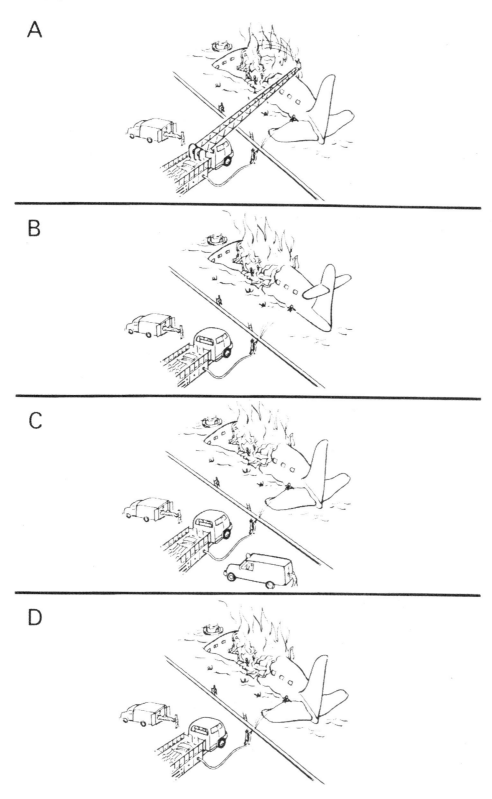

74. A flotation device can help a victim in a water emergency by giving them some _____ as they are pulled to safety.
    (A) boyancy
    (B) buoyancie
    (C) bouyanscy
    (D) boyencie
    (E) buoyancy

75. _____ tools often are used to remove victims trapped in an automobile.
    (A) Hidrolic
    (B) Hydrollic
    (C) Hydraulic
    (D) Hidroulic
    (E) Hydroulic

76. If there is an interior fire on an upper floor of a building, training water from the street and through windows from an aerial ladder might not be sufficient to extinguish the fire. You must bring water directly to the fire. Attach one end of a hose segment to the nearest available fire hydrant and carry that segment to the second-floor landing, unrolling as you climb. Meanwhile, other firefighters must carry rolled-up segments of hose to each upper staircase landing until the fire floor is reached. At the second-floor landing, the rolled-up segment is coupled to the segment connected to the fire hydrant and then is unrolled as firefighters carry it to the third-floor landing. The process is repeated until all hoses are connected to the fire floor. A firefighter at the street then turns on the water at the hydrant.

    Firefighters have arrived at a fire in progress in interior rooms on the fourth floor of a six-story building. Some have raised the aerial ladder and are directing streams of water through the fourth-floor windows. Other firefighters are wetting down the facade of the building. Firefighters are climbing from the fire engines carrying coils of hose into the building. Firefighter Poplis is holding one end of a partially unrolled hose in his hand on the sidewalk. The next thing Poplis should do is
    (A) turn on the water
    (B) attach the hose end to the hydrant
    (C) carry the hose end into the building
    (D) call out to ask if the hose has arrived on the fourth floor

77. Sometimes firefighters are called upon to help private homeowners who experience severe basement flooding. When arriving at a flooded basement, the first thing firefighters must do is sniff. If the water has an odor, firefighters must identify the odor as laundry overflow, storm sewer backup, or household sewage backup. If there is no odor, a search must be made for the source of the water. If firefighters can locate a separated or broken pipe, they can shut off the water supply behind the break. If the break cannot be found, firefighters can shut off the main entering the house from the street. They then can pump out the basement and advise the homeowners as to whom to call for repairs.

Firefighters Jonas and Barberi have responded to a homeowner's call for help with a flooded basement. They enter the basement to see water pouring through the ceiling. Jonas runs upstairs to the bathroom immediately above the waterfall and instantly recognizes that the toilet bowl is broken. He shouts his findings down to Barberi. Barbed should now
- (A) shut off the valve on the pipe leading to the bathroom
- (B) shut off the water to the house
- (C) begin pumping out the basement
- (D) tell the homeowner to call a plumber

DIRECTIONS: *Choose the appropriate word or phrase to complete the sentences in questions 78 through 80.*

78. Mrs. Austin was very thankful that the firefighters were able to save _____ house from the fire.
    - (A) his
    - (B) its
    - (C) her
    - (D) they
    - (E) them

79. _____ the firefighters liked it or not, all of the dirty hose needed to be cleaned and dried before it was put back on the truck.
    - (A) Although
    - (B) Whether
    - (C) When
    - (D) Whither
    - (E) While

80. The captain was in agreement _____ the other firefighters over the issue.
    - (A) to
    - (B) on
    - (C) of
    - (D) about
    - (E) with

81. While fighting a fire, a firefighter becomes trapped in a third-floor loft when the stairs catch fire. There is no fire escape. In this instance, it would be best to
    - (A) call for a portable ladder to escape through a window
    - (B) wait for help to arrive
    - (C) go to the nearest window and jump
    - (D) go up the stairs to the next floor and await a rescue attempt

82. After a fire has been extinguished, a firefighter observes a newly appointed firefighter pulling down loose plaster from a ceiling without a protective helmet on. It would be most appropriate in this instance for the experienced firefighter to
    - (A) advise the superior officer concerning the rule violation
    - (B) say nothing to anyone
    - (C) explain to the new firefighter the danger involved and request that he or she wear a helmet while performing this function
    - (D) go to any extreme, using force if necessary, to convince the new firefighter to wear a helmet

DIRECTIONS: *Answer questions 83 and 84 by choosing which implement in the lettered boxes usually is used with the object on the left. Note: The tools are not drawn to scale.*

83.

84.

85. As you are chopping holes into warm partitions in a house fire that appears to be fully extinguished, the homeowner asks what you are doing. Your best answer is that you are
   (A) chopping holes in the wall
   (B) making certain that the fire is not spreading inside the walls
   (C) following instructions
   (D) finding out why the walls feel warm

86. If you are instructing a new firefighter in proper placement of a 16-foot extension ladder, you should tell the firefighter
   (A) to lean the ladder on a firm wall
   (B) to place the base four feet from the wall
   (C) not to make the angle of the ladder too steep
   (D) to place the base a distance from the wall equal to one-fourth of the ladder's height

DIRECTIONS: *Use the following map to answer questions 87 through 89. The flow of traffic is indicated by the arrows. If only one arrow is shown, traffic flows only in the direction indicated by the arrow. If two arrows are shown, traffic flows in both directions. You must follow the flow of traffic.*

87. A patron at the municipal swimming pool is having an apparent heart attack. All emergency services crews are out on calls at this time, so the rescue team from the fire department must respond to the crisis. The best route for the rescue team to take from the fire station (which has exit doors at both east and west ends of the structure) is to
    (A) exit by way of 31st Street, taking 33rd Avenue to Rod Road, making a right, and taking Mott Street straight across to the pool
    (B) exit the east end of the fire station, turn left onto Rod Road and then left onto Broadway, proceed on Broadway to make a left onto 32nd Road, and then turn right onto Mott Street to the pool
    (C) exit onto Rod Road going north, take Skyline Drive westbound to 32nd Road, go four blocks, turn left on Pipeline, turn left onto Pennsylvania Avenue, and then turn left from Pennsylvania Avenue onto Mott Street
    (D) use the Moss Place exit of the fire station, go north one-half block to Broadway, take the long block of Broadway to a left onto 32nd Road and a right onto 33rd Avenue, turn south onto Sky Way, and turn right onto Mott Street

88. Two firefighters are performing a routine fire safety inspection at a store in the Blue Hill Mall. Their fire engine is parked just outside the Sky Way entrance to the parking lot. Their next assignment is to check a malfunctioning street box on the southeast corner of 32nd Road and 33rd Avenue. The most efficient way for them to get there is by
    (A) going south on Sky Way, going east on High Street to the corner of Rod Road, turning left onto Rod Road as far as Broadway, turning west on Broadway, and turning south on 32nd Road to its corner with 33rd Avenue
    (B) going around the parking lot to Arthur Avenue, going north onto 33rd Street, and then going west onto 33rd Avenue to the corner
    (C) going three-quarters of the way around the parking lot onto Arthur Avenue going west, taking Arthur Avenue to make a right turn onto Adams Street northbound, and making a right onto 33rd Avenue
    (D) continuing south on Sky Way to High Street, turning east for one block to make a left onto Pennsylvania Avenue, turning right, left, and left again to be headed west on Broadway, and turning left onto 32nd Road

89. Having repaired the street box, the firefighters are about to return to the fire station, but a road crew has just begun to repair a pothole on 33rd Avenue and has closed the street to all traffic between 33rd Street and 32nd Road. The firefighters will have to travel east on
    (A) Broadway
    (B) Pipeline
    (C) Mott Street
    (D) High Street

90. When arson is the suspected cause of a fire, the call must go out for witnesses. Each witness who comes forward must be interviewed to establish the credibility of the witness and the accuracy of the information. You first must identify the witness by asking his or her name, address, telephone number, place of business, and employer, if any. You then must question the witness as to how he or she happened to be in a position to witness the arson, and you establish the precise spot and angle from which the witnessing took place. You also must ask the witness to tell you exactly what he or she saw happen and ask for a complete description of the alleged arsonist—sex, approximate age, size, race, and clothing. If the alleged arsonist was known to the witness, you can ask for a name, address, and so on.

    As the last embers of a fire that totally destroyed a vacant storefront burn down, a somewhat disheveled elderly man approaches firefighter Kelliher and says, "I know who set that fire." Kelliher asks the man who he is, where he lives and works, and what he does for a living. Kelliher's next question should be
    (A) Why were you in this neighborhood this afternoon?
    (B) So, who set the fire?
    (C) Did you see the person strike a match?
    (D) Was the arsonist a man or a woman?

91. The public is warned never to use an elevator during a fire. Next to every elevator is a sign that reads, "In case of fire, use the stairs." Firefighters often use the elevators, however, to transport themselves and their equipment to upper floors more quickly. When firefighters enter a burning elevator building, they first must ascertain that no elevators are currently in operation and that no passengers are trapped in nonfunctioning elevators. If there are passengers on an elevator stuck between floors, firefighters must first remove these passengers. They should begin with an

attempt to restore power to the elevator and to bring it down to the ground floor. If the elevator cannot be moved by mechanical means, they should attempt to raise the elevator to the next higher floor by opening the elevator shaft and manually pulling the cables. If this maneuver fails, firefighters must go to the floor immediately above the level on which the elevator is stuck, open the elevator door with the firefighter key, and descend to the roof of the cab with ladders. They then can open the hatch of the elevator cab and enter the cab with their ladder. One firefighter must enter the cab to assist passengers with climbing up the ladder while two firefighters remain at the top, one to hold the ladder steady and the other to assist passengers as they alight. If there are no passengers trapped in elevators, firefighters must assess all safety factors. If water threatens the electrical system, firefighters must use the stairs.

Firefighters enter the 14-story building at 6 Livingston Plaza in response to a fire on the seventh floor. They check the elevator monitor board and note that, of the eight elevators in the building, two are in the lobby, one is on the fourteenth floor, one is on the third floor, one is on the sixth floor, and one is on the eleventh floor. One elevator is stuck between the fourth and fifth floors, and one is stuck between the ninth and tenth floors. The building staff has shut off the power to the elevators. Using an intercom connected directly to the elevator cabs, firefighters learn that both of the stuck cabs are unoccupied. Next, they should
(A) turn the power back on to get all cabs to the lobby
(B) attempt to manually hoist the stuck cabs to the floors above
(C) check the elevator electrical control box for water
(D) take the two elevators in the lobby to the seventh floor and fight the fire

92. At the scene of a building collapse, firefighters must search out victims and rescue them as quickly as possible. Search of a building collapse is especially hazardous because the area often is unstable. Debris might shift underfoot, and further material might fall on the victims and the rescuers. The procedure is to first rescue those victims who are easily accessible. Such victims usually can cooperate in their own rescue. Next to be rescued are victims who are visible but are either severely disabled or trapped into immobility. Great caution must be taken in removing the next class of victims, those who can be heard crying out but who are buried under rubble. When these victims have been brought to safety, firefighters must try to communicate with other buried victims by calling out or by banging on exposed pipes. Attempts to call victims must be followed by total silence so that distant, weak calls or taps can be heard. The search for victims must continue even after known survivors have been removed because it is possible that unconscious survivors remain buried by the collapse.

A large building under construction has collapsed on itself and has come down on many construction workers. Firefighters arrive and immediately begin plucking workers from scaffolding and from cables to which they are clinging. They then dig into the debris wherever they see flailing arms or legs or hear cries for help. Having carefully moved much rubble to extricate trapped workers, they call out "Is anybody else there?" Now they should
(A) go home because everyone else must be dead
(B) stop and listen
(C) go on digging cautiously
(D) put out the fire

DIRECTIONS: Answer questions 93 and 94 by selecting the paragraph in which the sentences appear in the most logical order and make the best sense.

93.
- (A) To maintain public confidence in that organization, even the appearance of corruption must be avoided. It is not permissible for fire inspectors to accept a meal "on the house" or to receive any other such favors or gifts. They must pay for their meals just like everyone else. The possibility of corruption or corrupt practices exists within every large organization.
- (B) It is not permissible for fire inspectors to accept a meal "on the house" or to receive any other such favors or gifts. To maintain public confidence in that organization, even the appearance of corruption must be avoided. They must pay for their meals just like everyone else. The possibility of corruption or corrupt practices exists within every large organization.
- (C) To maintain public confidence in that organization, even the appearance of corruption must be avoided. The possibility of corruption or corrupt practices exists within every large organization. It is not permissible for fire inspectors to accept a meal "on the house" or to receive any other such favors or gifts. They must pay for their meals just like everyone else.
- (D) The possibility of corruption or corrupt practices exists within every large organization. To maintain public confidence in that organization, even the appearance of corruption must be avoided. It is not permissible for fire inspectors to accept a meal "on the house" or to receive any other such favors or gifts. They must pay for their meals just like everyone else.

94.
- (A) Periodic fire drills are a very effective teaching device. Residents of a two-story house with only one staircase must have another way out of their second-floor bedrooms. Each member of the household must learn how to use the alternate route out. One solution is to purchase sturdy rope ladders in a hardware store and to bolt one securely to the underside of a windowsill in each bedroom.
- (B) Residents of a two-story house with only one staircase must have another way out of their second-floor bedrooms. Periodic fire drills are a very effective teaching device. One solution is to purchase sturdy rope ladders in a hardware store and to bolt one securely to the underside of a windowsill in each bedroom. Each member of the household must learn how to use the alternate route out.
- (C) Residents of a two-story house with only one staircase must have another way out of their second floor bedrooms. One solution is to purchase sturdy rope ladders in a hardware store and to bolt one securely to the underside of a window-sill in each bedroom. Each member of the household must learn how to use the alternate route out. Periodic fire drills are a very effective teaching device.
- (D) Each member of the household must learn how to use the alternate route out. One solution is to purchase sturdy rope ladders in a hardware store and to bolt one securely to the underside of a windowsill in each bedroom. Periodic fire drills are a very effective teaching device. Residents of a two-story house with only one staircase must have another way out of their second floor bedrooms.

95. In the event of a house fire, it is crucial for all residents of the house to be accounted for as soon as possible. In preparing your children for what to do if they must escape from a fire in their own home, the best instruction is to tell them to
    (A) meet under the tree directly across the street
    (B) come to where room is outside
    (C) meet in a corner of the yard
    (D) all get together right next to the fire hydrant

DIRECTIONS: *Questions 96 through 98 are based on the following forest fire. In this drawing, the scene is viewed by reinforcements running down a rise in the field toward the line of the fire.*

96. Which of the following drawings shows the same scene at the same time but from the vantage point of a firefighting helicopter hovering above the trench?

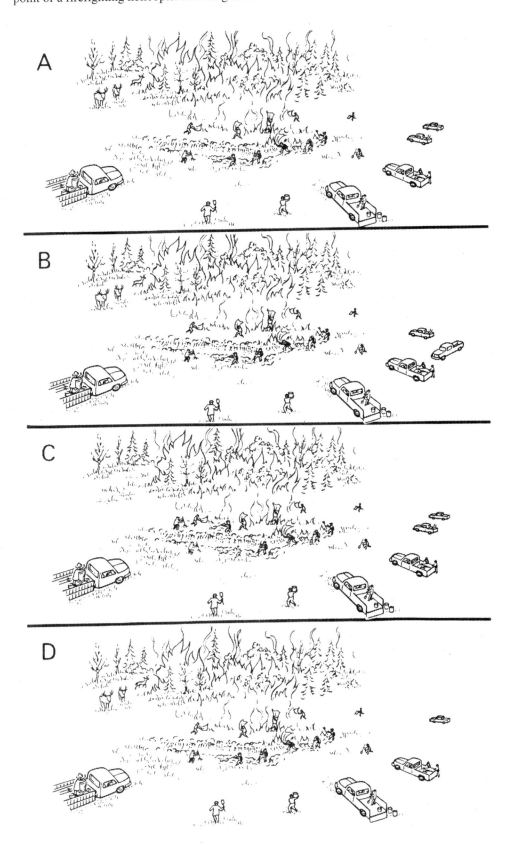

97. Which of the following drawings shows the same scene at the same time but as viewed from another helicopter approaching from over the flames?

98. Which of the following drawings shows the same scene at the same time but as viewed from a pickup truck filled with supplies parked at one end of the trench?

99. Firefighters must be called to a fire in a subway tunnel. The best instruction to give to help them locate the fire quickly and accurately is to tell them the fire is
(A) between the 59th and 66th Street stations
(B) about 300 yards from the 59th Street station
(C) in the southbound tube just north of the 59th Street station
(D) south of the 66th Street station, about 300 yards from 59th Street

100. A firefighter at the scene of a serious automobile accident sees that both the driver and the passenger are thoroughly wedged in by crushed metal. The firefighter will get the most effective assistance by broadcasting which one of the following messages?
(A) "I need help getting trapped passengers out of a car wreck."
(B) "Send heavy-duty cutting equipment at once."
(C) "There are injured people trapped in a car wreck."
(D) "The car is totally mangled."

Answer Key for Practice Examination 5

1.	D	21.	D	41.	C	61.	D	81.	A
2.	A	22.	D	42.	C	62.	D	82.	C
3.	B	23.	C	43.	D	63.	B	83.	A
4.	A	24.	C	44.	A	64.	C	84.	C
5.	D	25.	B	45.	B	65.	D	85.	B
6.	D	26.	C	46.	A	66.	A	86.	B
7.	A	27.	D	47.	D	67.	A	87.	D
8.	C	28.	A	48.	E	68.	C	88.	C
9.	D	29.	C	49.	B	69.	E	89.	B
10.	C	30.	B	50.	C	70.	A	90.	A
11.	D	31.	C	51.	D	71.	B	91.	C
12.	B	32.	B	52.	D	72.	C	92.	B
13.	B	33.	D	53.	B	73.	D	93.	D
14.	D	34.	A	54.	C	74.	E	94.	C
15.	D	35.	C	55.	A	75.	C	95.	A
16.	C	36.	B	56.	E	76.	B	96.	A
17.	A	37.	D	57.	A	77.	A	97.	B
18.	B	38.	B	58.	B	78.	C	98.	D
19.	D	39.	D	59.	B	79.	B	99.	C
20.	D	40.	E	60.	A	80.	E	100.	B

Explanatory Answers for Practice Examination 5

Confirm the correct answers to questions 1 through 18 by looking back at the sketch. If you made many errors, spend extra time on observation and memory training exercises.

19. **(D)** If you did not choose the correct answer to this question, reread the correct paragraph and notice the logical progression of the sentences and the even flow of the thoughts.

20. **(D)** See the preceding explanation.

21. **(D)** The purpose of first aid is to provide immediate medical attention until professional help arrives. You need to know the cause of the injury or disability only to the extent that it affects the first aid treatment. Thus, an overdose of pills dictates first aid steps, but the reasons for the overdose, bone breakage, or other disability are irrelevant to the first aid function.

22. **(D)** Although firefighters do act to save property, their primary purpose is to save people. The people must be kept out until it is safe for them to enter. Likewise, there is no reason for a firefighter to put him- or herself in jeopardy solely in the interest of property.

23. **(C)** The drill chuck on the left is used to grip the twist drill in box C.

24. **(C)** The drain on the left is unclogged with the plunger in box C.

25. **(B)** Because the captain has determined that there is no danger to residents and that no one needs to be evacuated, the next step is to put the fire out. The captain directs this effort.

26. **(C)** The fire engine must cross the sidewalk before it enters and turns into the street. As the situation is described, traffic in the street has been halted, but pedestrians must still be cleared and restrained before the fire engine pulls out.

27. **(D)** There might be more than one big nut, so (A) is not the best answer. In giving a direction to turn, the question arises: left or counterclockwise from the standpoint of the nut or from the standpoint of the person looking at it? The direction "to *your* left" makes this clear.

28. **(A)** Electric meters are large and are very readily identified. There might be more than one lower level to a building. What is a "far" corner? And most certainly there are two west corners—northwest and southwest.

29. **(C)**

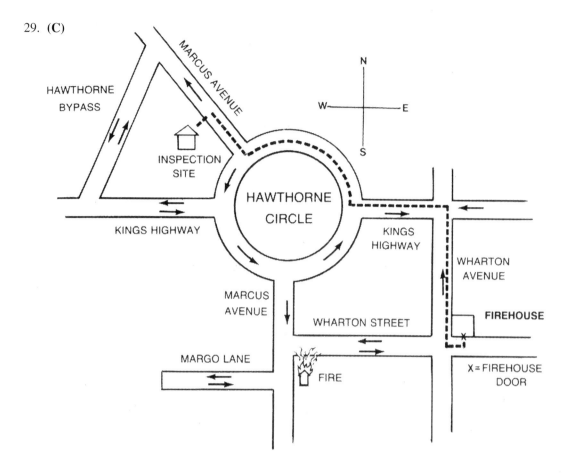

Try the other routes. (A) is wrong because you cannot exit directly onto Wharton Avenue; you must go west on Wharton Street and then turn right onto Wharton Avenue. Even if you could exit onto Wharton Avenue, Hawthorne Circle moves counterclockwise, not clockwise. (B) sends you north on the spur of Marcus Avenue that is one-way southbound. As for (D), Wharton Street will never get you to Hawthorne Circle.

30. **(B)**

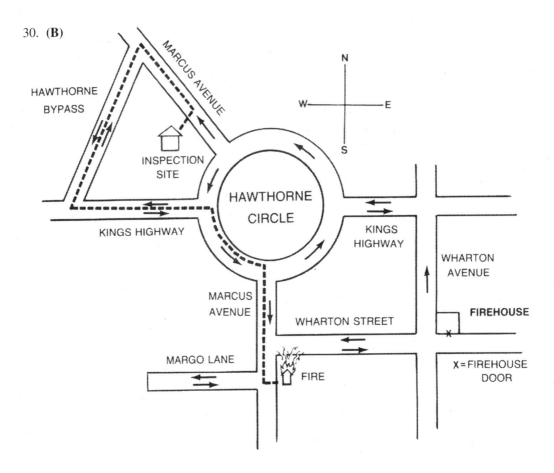

This spur of Marcus Avenue is one-way in a northwesterly direction, so (A) is wrong. As for (C), the firefighters must make a right turn from Kings Highway onto Hawthorne Circle, not a left. (D) presents the same problem as (A).

31. **(C)**

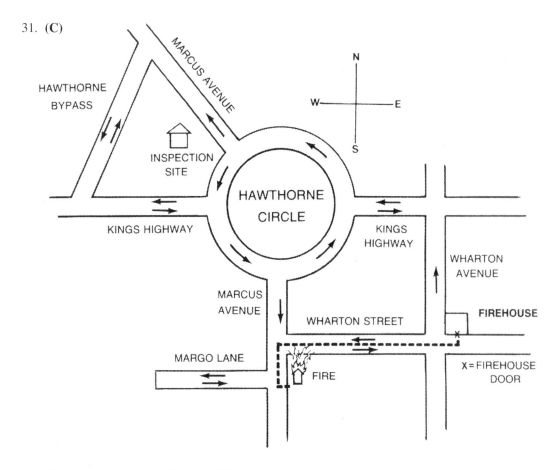

(A) cannot be correct because Wharton Street runs east-west. (B) is a legal route, but it is unnecessarily long and roundabout. (D) travels the wrong way around Hawthorne Circle.

32. **(B)** If your answer to question 32 does not agree with this, reread your choice and the correct paragraph once again. Subtle but real differences in continuity make the correct answer the best one.

33. **(D)** See the preceding explanation.

34. **(A)** This is a precise mirror image of the original drawing. In (B), the cat has jumped to the desk; in (C), ANTIQUES appears as read from the outside instead of reversed as from within; in (D), a large vase is missing.

35. **(C)** This is the window as seen obliquely from 17 feet above the street. In (A), the chair is on the wrong side of the desk; in (B), flowers have suddenly appeared in a vase; in (D), there is a different desk.

36. **(B)** This is the view from above. In (A), a new lamp has appeared; in (C), the bric-a-brac has been removed; in (D), the chair has been knocked over.

37. **(D)** The purpose of "tot finder" stickers is to identify the rooms in which small children sleep and are most likely to be found. Firefighters certainly should start their search as directed by the stickers.

38. **(B)** The odor of gasoline in a structure other than a gasoline station or garage arouses a strong suspicion of arson. Gasoline does not belong inside a house, office, store, or school. A "funny smell" is too vague, and the floor can be wet from water. Gas is used for heating and cooking. A smell of gas should not arouse suspicion of arson.

39. **(D)** The second paragraph states that firefighter Wilson checked the woman's vital signs.

40. **(E)** Engine 36 was dispatched at 11:32 PM and left the scene at 11:47 PM; thus, the call lasted 15 minutes.

41. **(C)** A firefighter who is not physically prepared to perform his or her functions can be the cause of much harm not only to him- or herself but to fellow firefighters and to the public.

42. **(C)** The building is on fire, and evacuation of occupants is the first concern.

43. **(D)** The drill in box D uses the drill bit on the left to make holes.

44. **(A)** The jumper cables on the left connect the car battery in box A to a source of power, usually another car battery, to start a car with a dead battery.

45. **(B)** The second paragraph states that the firefighters knew Ed was in the early stages of hypothermia and needed to be removed from the water as fast as possible. Therefore, after Ed was on shore, the firefighters were most concerned with treating his hypothermia.

46. **(A)** The second paragraph states that the firefighters knew Ed was in the early stages of hypothermia because he did not respond to their question.

47. **(D)** The firefighters slid the ladder along the ice for Ed to grab onto so they could then pull him out of the water. They did this because the ice was too thin for the firefighters to walk on.

48. **(E)** The second paragraph states that Ed had to be removed from the water as quickly as possible. His body temperature was dropping so rapidly that he soon would be unable to assist in his own rescue.

49. **(B)** The person calling to report a fire is likely to be very agitated. A specific question as to the location of the fire is likely to elicit the most specific answer. Even if the caller has the street number wrong, firefighters should have no trouble locating the fire once they arrive in the approximate neighborhood. A question such as "Where's the fire?" might elicit an answer such as "Next door." The answer to "Where are you?" might be "In the kitchen."

50. **(C)** The time of receipt of an alarm should be a matter of record. The question should be answered with the precise time.

51. **(D)**

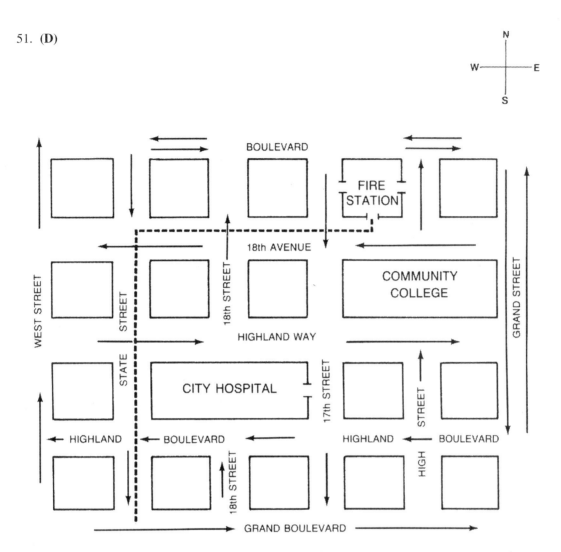

(A) cannot be correct because High Street does not intersect Grand Street. (B) confuses Highland Boulevard with Highland Way. (C) goes the wrong way on 17th Street.

52. **(D)**

Route (A) takes you the wrong way on the one-way 17th Street. (B) is a legal route but a long one; the injured firefighter should get to the hospital as quickly as possible. As for (C), Grand Boulevard does not intersect Highland Boulevard.

53. **(B)**

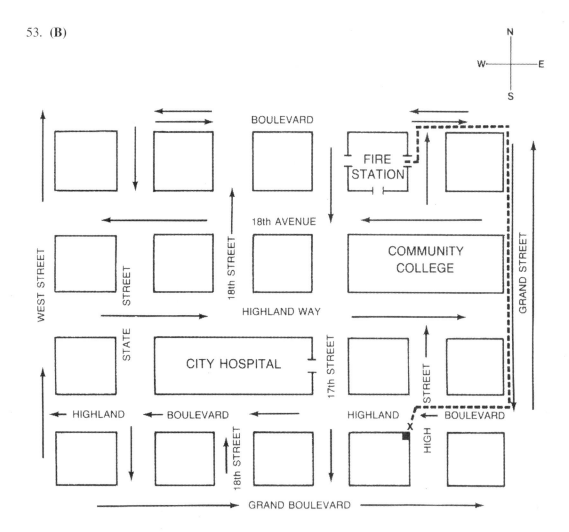

(A) is impossible for two reasons: High Street is one-way northbound, and Highland Boulevard is south of the fire station. Even if it were possible to go south on High Street, a straight run would be impossible because the Community College is two blocks wide and interrupts High Street. (C) is not correct because Highland Boulevard is one-way westbound, and High Street is east of 17th Street. (D) will not work because 17th Street alone will never get you to the corner of Highland Boulevard and High Street.

54. **(C)** The first paragraph indicates that the four essential elements of flaming combustion are temperature, fuel, oxygen, and chemical reaction.

55. **(A)** The first sentence of the second paragraph states that cooling with water is the most common way to extinguish a flaming fire.

56. **(E)** The first sentence of the passage states that flaming combustion can be stopped by removing at least one of the four essential elements.

57. **(A)** The last paragraph states that halons can be used to interrupt the chemical reaction that produces flaming.

58. **(B)** This drawing is accurate. (A) is incorrect because the life raft has been lowered all the way to the water. In (C), an extra fire has started on the barge. In (D), extra people have appeared on the barge.

59. **(B)** Obviously, from the opposite shore, everything is seen from its other side. In (A), the fire and smoke are blowing in the wrong direction. At the same time, the wind direction should be the same; only the viewpoint should change. (C) is incorrect because a structure has suddenly appeared on the barge. (D) is wrong because the tug has backed off.

60. **(A)** The view is from the opposite end of the scene. In (B), the life raft has been placed on the wrong side of the yacht. In (C), a person has appeared in the water and is being thrown a rope from the barge. (D) is incorrect because the fire has disappeared.

61. **(D)** In this instance, the job of the firefighters is clear. First, they must give first aid. Second, they should send for an ambulance, which will bring professional medical assistance. Finally, they should summon the police to investigate. The problem to be investigated is not a fire problem, so investigation is the job of the police, not the firefighters.

62. **(D)** Between answering alarms, firefighters spend much of their time making sure their equipment is functioning properly. Ineffective, malfunctioning equipment leads to ineffective firefighting and danger to the firefighters.

63. **(B)** The crowbar on the left is used to open the crate in box B.

64. **(C)** The mallet in box C is used to tap the chisel on the left.

65. **(D)** Because the total number of fires in 1995 was 1,043, the number of fires in Deveron in 1995 is equal to $1,043 - (374 + 9 + 465 + 132)$, or $1043 - 980 = 63$.

66. **(A)** The percentage decrease in Deveron is equal to $23/86 = 27\%$.

67. **(A)** For every four fires in Tolbury in 1994, there were three in 1995. In other words, $4 \times 3 = 12$ for 1994, and $3 \times 3 = 9$ for 1995.

68. **(C)** If the valve for hose A is turned on 3 seconds before hose B, it will have 3 seconds to fill at the rate of 5 feet per second. The water at this point travels $3 \times 5 = 15$ feet. This means that there is 110 feet of hose left to fill. When the valve to hose B is turned on, the water fills hose A at a rate of 2 feet per second. Because there are 110 feet of hose left, $110/2 = 55$ seconds. It took three seconds to fill the other 15 feet of hose. Thus, it took $55 + 3 = 58$ seconds to fill hose A.

69. **(E)** If 1 cubic foot of water can expand to 1,700 times its original volume, 4 cubic feet of water can expand to $4 \times 1,700 = 6,800$ cubic feet.

70. **(A)** The fire that was on the fourth floor is under control, and occupants of that floor and the floor above all have reached safety. The weight of water threatens those directly below on the third floor. They must be evacuated next.

71. **(B)** This is the view of the plane wreck as seen from a rescue launch approaching the nose of the plane. (A) is not correct because the nose is pointing too far up in the water. In the original, the plane has broken apart with nose down. (C) is not correct because the break in the plane is depicted behind the wings instead of in front. (D) is wrong because the waves are suddenly much too high.

72. **(C)** This is the scene from directly above. (A) is not correct because the firefighters with the hose are now working from a different angle. In (B), the wind has shifted, and the smoke and fire are now blowing in another direction. In (D), a rescue launch has turned and now is leaving the scene instead of approaching.

73. **(D)** This is the plane crash as seen from behind on the ground. In (A), an extra aerial ladder has suddenly appeared by the side of the plane near the break. The tail in (B) is not the tail of the original plane at all. (C) is not correct because the second ambulance was not in the original sketch.

74. **(E)** The correct spelling is "buoyancy."

75. **(C)** The correct spelling is "hydraulic."

76. **(B)** The hose is partially unrolled, so someone is already carrying it into the building and up to the second-floor landing. Clearly, this is the first segment of hose that is to be connected to the hydrant. Firefighter Poplis should connect the hose end to the hydrant. The connections must be secure at the hydrant and on every floor before the water is turned on.

77. **(A)** Because the source of the water has been clearly identified, firefighters need only turn off water to that area. There is no need to stop the water supply to the entire house. With the water off, they can pump out the basement and advise the homeowner to call a plumber to make repairs.

78. **(C)** The word "her" is the best word to complete the sentence. This is the proper pronoun in reference to Mrs. Austin.

79. **(B)** The word "whether" is the best word to complete the sentence.

80. **(E)** The word "with" is the best word to complete the sentence.

81. **(A)** The stairs are on fire, so the firefighter cannot use them to go either down or up. The question does not describe a situation so desperate that the firefighter should risk jumping from the third floor, but just waiting would be foolhardy. The firefighter should call for a ladder and go out the window.

82. **(C)** This is the perfect time for an experienced firefighter to share knowledge of safe practices with a rookie.

83. **(A)** The plane in box A is used to taper and smooth the board on the left.

84. **(C)** The trowel in box C is used to spread mortar between the bricks when building a wall.

85. **(B)** The homeowner is entitled to know why you are chopping holes in the walls when the fire appears to be out, so explain that fire can travel undetected inside the walls. Because the walls feel warm to the touch, you must double-check.

86. **(B)** Even a new firefighter does not need to be told to lean the ladder on a firm wall, but he or she might not know that proper placement of the feet of the ladder is a distance from the wall equal to one-fourth of the height of the ladder. With experience, this distance is easy to gauge, but you are advising a *new* firefighter. The height of the ladder is given as 16 feet, so tell the new firefighter to place the base four feet from the wall.

87. **(D)**

The answer cannot be (A) because there is no 31st Street exit and because Mott Street is not a through street. (B) ends up going east on a one-way westbound street. (C) would be legal and acceptable, but it is the long way around and thus is not the answer to "Which is the best route?"

88. **(C)**

(A) is both possible and legal, but it is not the most efficient route. (B) is short and efficient, but it ends up going west on a one-way eastbound street. (D) is another possible route, but it is longer than (C) and involves many turns.

89. **(B)**

With 33rd Avenue closed, the firefighters have no way of getting to Broadway or, for that matter, to High Street from their current location. Mott Street is accessible, but they cannot go east on Mott Street.

90. **(A)** According to the procedure and order of questioning set out, the firefighter must next establish the witness's purpose in being at that location at the time of the fire.

91. **(C)** No one is trapped inside the cabs stopped between floors, so the firefighters need not concern themselves with moving the cabs. There is no reason to call all the elevator cabs to the lobby. The firefighters must get to the seventh floor quickly to fight the fire, but before entering the elevator, they must ensure their own safety by checking the elevator shaft, control box, and wiring to rule out water damage.

92. **(B)** The next step, according to the instructions given here, is that the rescuers should stop and listen for a response to their calls and taps. If they hear any response, they must search for the responding victim. If there is no response, they must continue in a cautious search for unconscious victims. The situation described in the narrative makes no mention of fire.

93. **(D)** If you chose the wrong paragraph as the answer to this question, reread both your wrong answer and the correct one. The reason why the correct answer is correct should be quite clear.

94. **(C)** See the preceding explanation.

95. **(A)** Instructions for where to meet should be clear and unambiguous and should suggest a nearby spot out of harm's way. The yard has four corners, and mom could be almost anywhere outside, so neither of these is a good choice for a meeting place. Next to the hydrant is an unambiguous instruction, but it is a poor choice because of the likelihood of interfering with firefighting. The tree across the street is easy to locate, is close enough to allow for a quick count, and is out of the way of firefighters.

96. **(A)** This is the view from the helicopter. In (B), one of the pickup trucks is slightly turned. In (C), two diggers have moved between the people beating at the flames. In (D), a truck has disappeared.

97. **(B)** This is what the helicopter pilot would see. In (A), the line of beaters has disappeared. In (C), the flames are blowing down trench instead of across toward the firefighting crew. In (D), an airplane has arrived to join the effort.

98. **(D)** The correct drawing shows the same scene from the end of the trench. The escaping deer in (A) did not appear at all in the original drawing. In (B), a new fire truck, a pumper, has arrived, and the firefighters are spraying water on the fire. In (C), the flames are much higher than they were in the original.

99. **(C)** This is the only statement that clearly describes where the fire is—in which tube and at what location in the tube.

100. **(B)** The urgency of this message along with the call for heavy-duty cutting equipment makes it clear that people are trapped in crushed metal. Choice (D) never mentions people and thus does not imply urgency. (A) and (C) do not indicate what kind of help is needed.

APPENDIX

First Aid

Giving emergency medical care often is a part of the firefighter's role. Firefighters always have dealt with burns, broken bones, serious cuts and abrasions, and breathing stoppages. All of these medical emergencies occur with some frequency at the sites of major fires. Every new recruit to a fire department receives training in dealing with these medical emergencies as part of firefighter training. The training of firefighters is by no means restricted to training in putting out fires. In fact, in the current climate of cutting costs by combining administrative operations and by avoiding duplication of services, many municipalities are merging firefighting forces with emergency medical services. This means that some firefighters are being trained to serve as emergency medical technicians, and some emergency medical technicians and paramedics are being trained as firefighters.

No applicant about to take an open competitive exam for a firefighter position is expected to know how to drive a fire engine, how to raise an extension ladder, how to charge a hose, or how to immobilize a broken limb. Firefighter training academies offer training in all aspects of the firefighter's job. All recruits undergo rigorous and thorough training and must pass mastery examinations before being appointed to the force. Testing in knowledge and performance comes after instruction and practice. The initial open competitive exam, the one for which you are preparing right now in the hope of being accepted into the firefighter training program, will be testing your ability to learn, your common sense and judgment, and your general awareness of safety measures and first aid.

This introduction to basic first aid should help you answer judgment questions on your firefighter test and should provide you with some groundwork as you enter the firefighter academy for intensive, in-depth instruction.

HEARTBEAT

Your first thought with reference to injuries in connection with fires is undoubtedly that fires cause burns. This is reasonable. The firefighter is much less likely to have to deal with the burns, however, than to deal with breathing stoppages, heart stoppages, and shock. If a fire victim's heart has stopped beating, survival is limited to only a few minutes. A victim whose heart has stopped beating will have no pulse. Your training will have included cardiopulmonary resuscitation (CPR), and you must initiate CPR immediately—with one exception. If the fire is gaining on you and the victim, you both must get out. By remaining in the path of a fire to apply CPR to a victim, you simply create two victims. Pick up the victim and get out. This is a matter of common sense. You might encounter a question setting up a scenario such as this on your exam. The question might ask you which of a number of steps to take first in such a situation. The examiners are not really testing your knowledge of first-aid practices and procedures. They are looking for signs of logical reasoning and good judgment on your part. Removing the victim from the fire is ALWAYS the first step.

BREATHING

If your fire victim's heart is beating, the next priority for your attention is the victim's breathing, regardless of any other injuries he or she might have sustained. Check for absence of breathing by observing the victim's chest or by putting your ear to his or her nose to listen or feel for signs of breathing. An individual can live for only a very short time without breathing. A person who manages to survive after a period without breathing is apt to suffer severe brain damage. The preferred method of attempting to restore breathing is mouth-to-mouth resuscitation. This is a relatively simple procedure.

1. Position the victim on his or her back.
2. Clear the upper airway by turning the head to one side and running your finger through the mouth to remove the tongue from the throat and to wipe out debris.
3. Tilt the head back so the neck is stretched and the head is in a chin-up position.
4. If the victim is an adult, pinch the nose closed. If it's a child, you should cover both the nose and mouth with your own mouth.
5. Take a deep breath, open your mouth wide, and breathe forcefully into the victim's mouth. Watch to see if the victim's chest rises. If not, breathe harder. If your breath is still not reaching the victim's lungs, there might be an obstruction in the airway. Turn the victim and strike him or her between the shoulders to dislodge the obstruction.
6. Take another deep breath and again breathe into the victim's mouth. Repeat rhythmically 12 to 20 times per minute until victim breathes alone or until help arrives.
7. With infants and small children, breathe more shallowly from your cheeks rather than from deep in your lungs.

Your examination will not ask you how to perform mouth-to-mouth resuscitation, but it might ask you a question about when such a procedure is appropriate or how long you should continue. Your answer will indicate awareness and judgment.

BLEEDING

Severe bleeding is not a direct result of exposure to fire. Nonetheless, severe bleeding among fire victims is not uncommon. The bleeding is, of course, the result of an injury. A person attempting to escape a fire might rush through a smoke-filled area with poor visibility and crash through jagged glass or impale him- or herself on fire-damaged wood or metal. The escape attempt might entail a fall or a jump that leads to an injury with severe bleeding. Bleeding often looks worse than it is. Spurting bright-red blood from a severed artery, however, leads to massive loss of blood and often is fatal. Uncontrolled, long-term heavy bleeding (hemorrhage) from a large vein also can prove fatal.

The person bleeding heavily should first lie down. If the bleeding is from a limb, the limb should be elevated to slow the blood flow. The firefighter, EMT, or bystander pressed into service should then cover the wound with the cleanest cloth readily available and press hard. Pressure over the wound should slow the bleeding, thereby limiting the loss of blood and allowing formation of a clot. You might have to maintain this pressure over the wound for a considerable period of time as you wait for a self-limiting clot to form. You will know that a clot has formed when bleeding has noticeably slowed; the bloody area of the covering cloth will stop expanding, and perhaps the cloth will feel a bit drier. You then can secure the cloth over the wound by tying it snugly but not tightly. If pressure over the wound stops the bleeding, it definitely is the treatment of choice. Because maintaining pressure on a wound can entail a considerable length of time, you should look around for an onlooker who can take over this task so that you, the professional firefighter, can return to the fire scene to search for more victims or to help fight the fire. Recognizing when and how to utilize nonprofessional help also is a part of judgment.

Arterial bleeding (the bright-red spurting kind) sometimes can be stopped, at least temporarily or in conjunction with direct pressure over the wound, by applying pressure on pressure points. Pressure points are areas on the body where the artery lies close to bone; external pressure can narrow the artery against resistance from the bone. There are pressure points at the temple just in front of the ear, below the jaw, behind the collar bone, in the upper and lower arms, at the wrist, in the groin, at the knee, and in the ankle. Apply pressure at the pressure point between the wound and the heart that is closest to the wound. If you apply pressure at both the appropriate pressure point and over the wound itself, you can slow the flow of blood and hasten clot formation. The diagram on page 366 shows the location of pressure points on the body.

A tourniquet is a tight, constricting band placed entirely around an extremity (arm or leg) to totally stop the flow of blood. A tourniquet is truly the method of last resort. A tourniquet shuts off the entire flow of blood to all tissue beyond it. Lack of blood causes tissue to die. After a tourniquet has been put into place and tightened, it must not be loosened. A tourniquet should be removed only by a doctor. Before placing a tourniquet, consider how soon the victim can reach qualified medical attention. On the other hand, if you find it impossible to stop arterial bleeding in any other way, it is surely better to lose the limb and save the life. This is a judgment call. Delaying a tourniquet and letting the victim bleed to death would not be prudent.

Your firefighter exam will not ask you to name pressure points, though it surely can do you no harm to know where they are, but it might ask a general question about control of bleeding to check your grasp of priorities.

OTHER EMERGENCIES

Absence of heartbeat, breathing stoppage, and hemorrhage represent three emergency situations in which firefighters must take immediate, positive action. In these situations, time is of the essence. If you find yourself facing any one of these medical emergencies, you can not wait for an EMT, a paramedic, an ambulance, or a doctor. You must act. And you must remember this for answering questions on your firefighter exam.

Another group of emergency medical situations require immediate attention from firefighters and EMTs at the scene but do not require quite such instantaneous or proactive response. These situations, while not immediately life-threatening, can lead too easily to shock and subsequent death. Two of these medical events, in particular, are fractures and burns.

Pressure points for temporary control of arterial bleeding.

Fractures

A fracture is a break in a bone. The broken bone can be in an extremity such as an arm, leg, hand, or foot; in the head such as a skull fracture; or in the torso such as a broken collarbone, rib, or pelvis. Fractures must be diagnosed and treated by a physician, and they are extremely painful. A fracture victim in extreme pain may well go into shock. (Shock is discussed in detail later in this appendix.) The role of the firefighter at the scene is to lessen the chance or severity of shock by alleviating pain. The pain of a fracture is best alleviated by immobilizing the affected area. If it is safe to leave the victim wherever he or she is sitting or lying, the person should be told to stay in one place and move as little as possible. If the person must be moved, movement must be done as gently as possible with minimal jarring to the broken bones. Where possible, a makeshift splint should be fashioned, extending beyond joints on either side of the break. Such a splint—made of a board or stick, padded with cloths or grass or moss, and secured with string or belts—should keep pieces of bone from moving and thereby spare flesh from friction. The less motion, the less damage to tissue, the less bleeding, the less pain, and the less likelihood of shock.

In cases of fire, bone breakage often comes from jumps from high windows and from falls through floors weakened by the fire. Unfortunately, these jumps and falls frequently cause back injuries, and the fracture with the greatest potential for grave damage is a fracture of the spine. Enclosed within the bony structure of the spinal column is the spinal cord. The spinal cord carries nerve messages from the brain to the nervous system and hence to the muscles of the entire body. If the spinal column is severed, the messages cannot get through. Broken bones have sharp, ragged edges. A broken bone in the spinal column, if moved, can injure the spinal cord. A spinal cord injury in the lower spine affects the lower extremities. It can leave the victim a paraplegic, unable to walk. A spinal cord injury farther up on the spine can affect all parts of the body below the injury. A broken neck can leave the victim paralyzed from the neck down, a quadriplegic. The general rule in first aid is that a person with a suspected or even possible back or neck injury should never be moved until totally immobilized on a back board. In firefighting, however, there is always the one consideration that takes precedence—getting away from the fire. If the choice is between possible paralysis and certain death, is there a choice? Remember this as you take your firefighter exam.

Burns

The most obvious effects of exposure to fire are smoke inhalation and burns. Smoke inhalation does not lend itself to first aid. The firefighter can only administer oxygen as instructed at the training academy and wait for professional medical assistance. The burns of the burn victim also cannot be treated at the scene, but the burn victim can and must receive attention. Except for the most superficial burns, all burns are serious. Body fluids are lost very rapidly through the large expanses of exposed flesh of burns, and this can lead to dehydration, lowered blood pressure, and shock. The intense pain of burns also is a contributor to shock. After removing the burned person from the fire, the firefighter must make that person lie down. The firefighter then should loosely cover burned areas with the cleanest cloth available. If the person is conscious and is not extremely nauseated, the firefighter must encourage him or her to begin fluid replacement as soon as possible. Lightly salted cool or cold water is best, but if salt is unavailable, plain water will do. The burn victim should start with small sips and then work up to full glasses.

Shock

Shock is the body's involuntary response to major injury. Shock can occur immediately upon the injury to the body or might develop as a reaction later. Shock is characterized by a swift and severe loss of blood pressure, rapid lowering of body temperature, rapid heartbeat, bodily weakness, and frequently, fainting. With lowered blood pressure comes decreased blood circulation and oxygen starvation of cells, especially cells of the vital organs. In particular, the brain, liver, heart, and kidneys can be rapidly damaged by shock. If shock deepens and continues, the result is liver and kidney failure.

The injuries that fire victims sustain are precisely the types of injuries that tend to throw a person into shock. For a fragile personality, fear, fatigue, and pain are adequate to initiate the shock syndrome. In combination with loss of blood from hemorrhage or loss of body fluids from burns, the fear, fatigue, and pain almost guarantee shock. In addition, fire victims often face legitimate apprehension about the fate of their loved ones who might have been involved in the same fire and insecurity about their future health, homes, and lifestyles. The physical and mental pressures on a fire victim are almost overwhelming. Shock is the response.

Everyone who deals with fire victims must be aware of symptoms of shock and of the best ways to forestall, lessen, or reverse these symptoms. The medical treatment of choice is intravenous replacement of lost fluids—blood or plasma by transfusion and other fluids by infusion of saline, sugar, or electrolyte solutions. Obviously, this medical treatment must follow the first-aid treatment by the firefighter at the scene. The firefighter's role is to:

1. Remove the victim to a safe location for the safety of the victim as well as to allay the victim's fear of danger and additional pain that might be caused by the fire.
2. Convince the victim to lie down. Unless the injuries themselves keep the victim off his or her feet, a person who has just been traumatized by fire is likely to be highly agitated and to walk about frantically. This type of activity will hasten shock. You must calm the victim.
3. Control bleeding. Control of bleeding is a lifesaving measure in itself. Blood is absolutely necessary to sustain life, and massive loss of blood is seriously life threatening. More moderate loss of blood also is contributory to shock. In combination with loss of blood pressure and diminished circulation, lower blood volume can lead to a critical situation. As a firefighter, you can contribute to shock prevention by helping to stop the bleeding and by encouraging the victim to drink water to replace lost fluids and prevent dehydration.
4. Make the victim comfortable. A common reaction to trauma is shivering. If the victim's clothing is wet, remove it. Loosen constricting clothing. Cover the victim loosely. Keep him or her warm but avoid overheating. Extreme chilling and extreme overheating both adversely affect blood circulation. You must watch for erratic changes in temperature and adjust the coverings accordingly. If a cold victim is not nauseated, warm liquids can be soothing and can help with the warming process. A nauseated victim, on the other hand, should be served cool or cold liquids to avoid vomiting.

A few words about liquids:

- Never try to force liquids into an unconscious person.
- If the victim is unable to sit up, administer liquids with a straw or spoon. A person who cannot lift his or her head to drink might be able to get liquids by sucking on a cloth repeatedly dipped into the liquid. This method also might work for a person who is nauseated. It is less likely to induce vomiting.
- Never administer alcoholic drinks.

5. Relieve pain. The firefighter at the scene has very limited resources for relieving pain. Immobilizing broken bones can help. Covering open wounds to protect them from wind, water, or falling debris also can be useful. Giving reassurance and moral support can help, too. Believe it or not, your calming words can help slow the victim's breathing and heart rate. You can make the victim feel better and can slow the progress of shock just by talking gently.

In short, as a firefighter, you are the first at the scene of a fire. As such, you face the primary mandate to save lives. You save those lives by removing people from danger caused by fire itself, by smoke, and by crumbling structures. If any people you remove from danger require immediate lifesaving attention, it is your responsibility to initiate that care until some other person arrives to take over. Your task then is to extinguish the fire.

These are your priorities. First save lives and then save property. First help keep victims alive and then do what you can to reduce physical damage to those victims. If you remember these simple rules, you should be able to answer most of the common sense and judgment questions on your firefighter entrance exam.

VEHICLE SAFETY

Mechanical breakdown of firefighting vehicles is absolutely unacceptable. To avoid the possibility of breakdown, all fire vehicles are subject to regular maintenance and frequent overhaul. Maintenance of vehicles is performed by trained, qualified automotive maintenance mechanics. Firefighters do not change belts and hoses nor do they do tune-ups and wheel alignments. On the other hand, firefighters are in daily contact with their vehicles and are in the best position to notice potential problems. If you recognize or suspect a vehicle safety problem, you are obligated to bring this problem to the attention of your supervisor or a mechanic or to follow your procedure manual as to the line of reporting.

Very little of the firefighter's time is spent actually fighting fires. A large part of every day is spent in physical fitness activities, in training, in making fire inspections, and in preparing vehicles to fight fires. Vehicle preparation includes making sure all equipment carried by the vehicles is in place and is in good condition. Hoses must be correctly coiled and must not leak. Airpacks must be in place along with axes, pikes, and ropes. Equipment must be secured so it remains in place while the vehicle is in motion but so it can be easily released for rapid deployment at the fire scene. Every day, you will check and recheck to be sure everything is ready. You also will clean and polish the vehicles with pride.

As long as you are spending so much time clambering on and around the vehicles, checking equipment, and polishing the brass, you might as well keep alert to the condition of the vehicles themselves. Do not limit your attention to the big pumpers and hook-and-ladders. The smaller trucks that go out on inspections, the ambulances, and the chiefs' cars all should be in perfect condition at all times.

First of all, look. Look under the vehicle. Is there a puddle? Use your eyes and your nose. A gasoline leak will be close to the gasoline tank and will smell distinctly like gasoline. Antifreeze/coolant is green. Motor oil is brown and is thick and greasy. Transmission fluid also is oily but is pinkish and is a thinner oil.

After looking for puddles, look at wheels and tires. Be sure all tires are properly inflated. Mechanics should be monitoring tread depth for you, but you can double-check. Also look at the tire sidewalls. Steel cords or fabric popping out of the sidewalls is very dangerous. Be sure all lugs are in place and tight and that rims are not bent or damaged. Take a look at valve stems to be sure they are in good condition. If even one tire leaks, the driver will find maneuvering more difficult.

Don't crawl under the vehicle, but take a look at the tail pipes and other visible parts of the exhaust system to make sure pipes are not missing, broken, or compressed.

As you walk around the vehicle checking for leaks and tire condition, notice the lights. If the glass is dirty, wash the headlamps, taillights, signals, flashers, reflectors, and so on. You need to have clean lights to see and to be seen. This is such a simple safety measure, yet it really pays off.

As long as you are wiping or washing off the lights, hop up and do the same for mirrors and the windshield. Obviously, greater visibility leads to greater safety. While looking at the windshield, check the wiper blades to be certain that they are secure and that they make contact with the windshield throughout their length.

As you walk around the vehicle, you might notice body damage, hairline cracks, or tiny holes in the windshield or other glass. Tiny breaks or holes in the windshield should be repaired before they become so large that the whole piece of glass needs to be replaced. Bring small problems such as these to the attention of the supervisor or mechanic so they can be attended to at the next regularly scheduled overhaul or even sooner if necessary.

Now get into the driver's seat. Try the horns, bells, and sirens. Turn on lights one-by-one and make sure all of them work. You might need someone to stand behind the vehicle to tell you whether your brake lights and backup lights are working properly. Turn on the windshield wipers and try the washers. Everything should work. Start the engine and listen for strange noises. Look at all the gages and warning lights. Some of these might not be operative until the engine has been running for awhile or until the vehicle is in motion, but watch all those that might indicate problems in a stationary vehicle. With the engine running, you can test the brakes for pressure even if you can't test them for stopping speed or balance. You also should check the steering wheel to be sure it does not have too much play. You want the steering wheel to be responsive when the driver turns it. Finally, open and shut all doors. Make sure they latch securely.

Unless your departmental regulations tell you otherwise, the next place to look is under the hood or in the engine compartment. There are many safety checks you can make here without any formal mechanical training. First, check the fluid levels. The fluid levels you can easily check on cars and on some other vehicles are:

- Motor oil
- Transmission fluid
- Brake fluid
- Power steering fluid
- Radiator coolant
- Windshield washer fluid
- Battery fluid

Next look at belts and hoses. Belts and hoses must be securely attached and must be neither stiff and rigid nor soft and spongy. There must be no nicks or cuts that could lead to breakage on the road. If you are not certain as to just how belts and hoses should feel, ask a supervisor or a mechanic. There is a very distinct feel to dried-out rubber and to rotting rubber. After you have been shown, you will know how firm but flexible belts and hoses feel.

While checking in the engine compartment, you should look for loose electrical connections, frayed wires, or buildup of deposits of any sort. If you notice any wires or parts that seem wobbly or out of line, ask for a second opinion. You are not expected to be an expert, but you are expected to be alert to potential problems or safety hazards.

Firefighting is team work. You will never find yourself alone at the scene of a fire. Likewise, you will never be the only person looking after the vehicles in the fire house. All maintenance activities will be carried on with the advice and cooperation of your fellow firefighters.

On the other hand, you are on your own when you sit down to take the firefighter entrance exam. Think about the aspects of vehicle safety that might be your responsibility as you read the exam questions. Again, remember that you are not expected to have any specific knowledge about fire vehicles. You are expected to have common sense and to know, for example, that greater visibility leads to greater safety. So give common sense answers and improve your chances for selection.

Notes

Notes

Notes

Notes